THE ECHOING UNIVERSE

Also by Emma Chapman

First Light: Switching on Stars at the Dawn of Time

THE ECHOING UNIVERSE

HOW RADIO ASTRONOMY HELPS US SEE THE INVISIBLE COSMOS

EMMA CHAPMAN

BASIC BOOKS
New York

Basic Books
Hachette Book Group
1290 Avenue of the Americas, New York, NY 10104
www.basicbooks.com

Printed in the United States of America

Originally published in 2026 by John Murray in Great Britain
(as *The Radio Universe*)
First US Edition: May 2026

Published by Basic Books, an imprint of Hachette Book Group, Inc. The Basic Books name and logo is a trademark of the Hachette Book Group.

Library of Congress Control Number: 2026934920

ISBNs: 9781541601857 (hardcover), 9781541601871 (ebook)

LSC-C

Printing 1, 2026

To my dad,
who woke me up to see the comet,

and to my mum,
who made sure I was wrapped up warm.

CONTENTS

The most important feature of any telescope is the imagination with which it is used.

Vera Rubin, 'The Dedication of the Vatican Telescope', 1993

INTRODUCTION

Whenever you look above the horizon, whether in appreciation of a sunrise or lost in the depths of a starry night, you step into the role of an astronomer. Look up at the Moon and your eyes transform into telescopes, as the pupils widen to gather its faint glow and the lenses focus the light onto the retinas. You are left contemplating an image that has guided sailors, aided farmers and enchanted poets and lovers for millennia. In various early societies, stargazing could determine a person's fate. Those who could spot more than five stars in the 'Seven Sisters' cluster were destined to be warriors, their sharp vision prized for spotting enemies and striking with precision. But even those of us with perfect vision can only see so much. The Moon is close enough for us to see its shadowy plains, but even the nearest planets are little more than bright pinpricks in the sky, their valleys and volcanoes hidden beyond the reach of our sight. And as for other galaxies . . . the luckiest of us might glimpse a faint smudge of light in the direction of Andromeda, but the shape of its spiral arms is entirely lost to us. To look further, we need an 'eye' that not only collects and focuses light but magnifies it too.

You can probably picture an optical telescope, even if you have never looked through one. At its most basic, it consists of a set of lenses or mirrors shaped to collect incoming light and focus it upon a magnifying lens in the eyepiece. The apparatus is usually housed in a metal frame mounted on a tripod that allows the telescope to revolve and tilt. When Galileo Galilei used one of his novel telescopes in the early 1600s, he was able to see details invisible to all humans before him, such as the moons of Jupiter. The shape of an optical telescope is so unchanged that Galileo would easily recognise the amateur telescopes set up in back gardens worldwide. He might marvel at the most modern, 'smart' versions which dispense with the eyepiece and instead use cameras to capture and send images to an app, but the physics and engineering of collecting celestial light remain the same.

When Galileo drew the movement of Jupiter's moons or charted the stars making up the ribbon of the Milky Way, his sketches were purely of academic interest, seen only by a privileged few. Now, astronomy is a global fascination, and images from the world's most advanced telescopes are shared across headlines and social media.

Almost all the astronomy images in the public consciousness have been captured from just one type of light: optical light. This light requires no special processing or translation, beyond magnification. In the fruit bowl of astronomy, optical light is the apple. It is light our eyes have evolved to see and, like an apple, there is no preparation needed in order to enjoy it. We simply bite into it and savour the taste. My own house is full of wall art captured by the Hubble Space Telescope; its beauty is immediate and undeniable. But what a waste to ignore the potential of the more exotic, elusive offering that lies next to the apple! Radio light is a 'fruit' that cannot be enjoyed so easily, the lemon of the bowl.

Eating lemons requires tools and ingredients: a knife, a juicer, extra sugar. But what versatility, what promise! There is the cake I could bake by adding a generous grating of zest in the batter. There is the soothing hot drink I could concoct, balancing the fruit's natural bitterness with a teaspoon of honey. And, being a scientist, I cannot help but look at a lemon and think of the secrets I could write, squeezing the juice to use as ink and watching my words disappear as it dries. Once heated by an iron, the carbon bonds in the citrus ink break down, darkening the juice and causing the letters to re-emerge miraculously on the page.

While optical astronomy is open to anyone who can look up, radio astronomy uses light invisible to our eyes and requires some extra tools: an antenna, a dish. Like lemon-juice ink, the mysteries of radio light remain hidden until we learn how to reveal them – but when we do, a cosmos of forgotten echoes and newfound wonders emerges. From mapping lunar craters with unparalleled resolution to filling in the timeline of the Universe, from listening to the sweeping screeches of Jupiter's Northern Lights to hearing the 'ping' of meteors streaking through the sky, from searching for extraterrestrials to spotting incoming asteroids, no type of astronomy matches radio for its breadth of application. Radio astronomy illuminates secrets that have been written in space over billions of years.

First Contact with the Radio Universe

Optical astronomy is as ancient as civilisation itself, born from the simple fact that we arrive into this world already equipped to see starlight. Radio astronomy, in contrast, began in 1928, with one man getting annoyed at the hiss in his headphones. Karl Guthe Jansky, an American radio engineer fresh out of college, had taken a job at one of the world's greatest scientific development companies, Bell Telephone Laboratories in Holmdel, New Jersey. The company had a problem.

By the late 1920s transatlantic telephone calls had become possible, with Bell Labs charging sums so steep that a single minute could cost nearly a week's wages for the average worker. To get away with charging that much, the company needed their phone lines to be crystal-clear.

Jansky had been tasked with characterising the crackling static that plagued the telephone lines, often making conversations difficult to hear. There were multiple culprits and many, such as distant electrical equipment, could be readily identified. But not all of them. Among the hums and crackles was a persistent hiss, one that defied explanation. He got to work immediately, and after three years he had a radio antenna capable of isolating the radio static. It wasn't a slight affair, at 30 m wide and 6 m tall, or an elegant one either: a tangle of wood and wire. He fondly referred to it as the 'merry-go-round', and go round it did, on a turntable set on four salvaged Ford Model T wheels.

For over a year, Jansky wheeled the antenna around, tracking the signal as it appeared to move across the sky, from east to west, eventually realising that it must be coming from deep within the Sagittarius constellation. This was a big surprise because, up until that point, the whole Universe was thought to be silent in the radio wavelengths. The *New York Times* devoted an entire front-page column to the discovery, alongside articles about Nazis parading through the streets of Germany. Perhaps in those years, as the storm clouds of war gathered on the horizon, there was a welcome distraction in looking up. Now, we know that Sagittarius is not just any constellation. It contains the heart of the Milky Way, and Jansky had been listening to radio light emitted by matter falling helplessly towards a supermassive black hole situated there.

Jansky had been given a simple task: to make our conversations clearer across continents. This he did – but, with the help of his merry-go-round of wire and wood, he also unwittingly opened a line of communication for astronomers. Now they could converse with the cosmos far beyond the confines of visible light – the only language astronomers had known until then.

A Tour of the Electromagnetic Spectrum

Light and its associated terminology permeate our daily language. There is light at the end of the tunnel as we wrap up a difficult project, the smile of a close friend lights up a room, and new information 'sheds light' on a situation. We seek clarity by asking someone to 'enlighten' us, having only had a 'glimmer' of understanding. Light represents knowledge, warmth, truth and safety. It banishes the darkness we fear on a primal level.

However, the physical nature of light perplexed scientists for centuries, with our understanding oscillating between theories that described it as a wave or a particle. Evidence piled up in favour of both sides, seemingly at odds. Only with the advent of quantum mechanics in the early twentieth century did physicists realise that the best description would change depending on the experiment.

In the quantum world, light experiences *wave-particle duality*. In experiments involving the refraction or bending of light, it acts like a wave. But we can only make sense of how light transfers energy by considering it to be made of particles – called *photons*. These photons can be thought of as particles with wave-like properties. I usually refer to radio *waves* because most astronomical phenomena are best described by light as a wave, but radio *photons* is just as valid a description in certain situations.

Whether we think of light as a wave or a particle, its properties vary according to its place on the *electromagnetic spectrum*, the range of possible values for the wavelength of the wave, or energy of the photon. The higher the energy of the photon, the shorter the wavelength of the wave. Imagine dangling your feet into a swimming pool. If you lazily move your feet back and forth, the ripples outwards will be spaced quite far apart. If, however, you kick your feet rapidly, the spacing between waves – the wavelength – will be shorter.

Let us return to our fruit bowl. Though the term 'fruit' describes everything in the bowl, I don't expect every lemon, orange and pear, or even every apple, to taste alike. In the same way, the different types of light making up the electromagnetic spectrum, such as radio and optical light, display very different properties. Light's spectrum ranges continuously from longer wavelengths in the order of kilometres (lower energies) to shorter wavelengths far, far smaller than a human hair, or even an atom (higher energies). This spectrum is divided, quite arbitrarily, into seven categories: gamma (shortest wavelength), X-ray, ultraviolet, optical, infrared, microwave and radio (longest wavelength).

There is a remarkable range of properties dependent on wavelength, even within one 'type' of light. In the optical range, we call the different wavelengths 'colours', and even within each one, our eyes can perceive a seemingly endless array of hues. Who hasn't stood paralysed in a hardware store, staring blankly at swatches of 'Overtly Olive' and 'Sophisticated Sage', trying to decide which shade of green would best suit the kitchen? Some may groan, 'They're the same colour!' But thanks to the three types of cone cells in our retinas, the average human can differentiate up to one million different hues, even if our patience gives out long before our eyes do.

Different wavelengths also help our bodies to self-regulate. Because blue light is more prominent during the daytime, our bodies respond to a decline in blue light by producing melatonin – the hormone that makes us feel sleepy. This is why looking at your phone in bed makes it more difficult to get to sleep. When your eyes detect the blue light from a screen, your brain assumes the Sun is still high in the sky and hurriedly delays the production of melatonin in order to keep you awake. And when your body finally does release melatonin, it does so well into the night, resulting in grogginess in the morning. On the International Space Station, astronauts experience sixteen sunrises and sixteen sunsets every single day, as they orbit the Earth every ninety minutes. Trying to adapt to that schedule would play havoc with the human body, so NASA has designed the station's lighting to mimic the day–night cycle on Earth. In the evenings, the blue light is reduced to help the astronauts sleep.

Every day, we encounter light that either helps or hinders us. We are so used to the concept of light being visible that we may forget we are also talking about light when we refer to microwaves or X-rays. The familiar rainbow of visible colours represents the full spectrum of the optical wavelengths, from red, the longest, through orange, yellow, green, blue and indigo to violet, the shortest. However, this is a tiny sliver of the whole electromagnetic spectrum.

Shorter than violet wavelengths are ultraviolet (UV) wavelengths, which are too short to activate the cone cells in our eyes and so invisible to humans.* This kind of light has enough energy to penetrate the surface layers of our skin, causing burns in the short term. The energy of UV light can also knock electrons away from the atoms that make up our cells, a process called *ionisation*. Because DNA is built from those same atoms, ionisation directly alters its structure, leading to long-term cellular damage in the form of premature ageing. An unwelcome attribute, one might think, except that these wavelengths also damage the DNA in bacteria and viruses, preventing them from replicating. Hospitals use UV sterilisation extensively, as do water purification centres, so as much as putting on sun cream feels like a pain, just be thankful pathogens have not evolved to excrete a natural SPF either.

As we move to higher energies, we come to the X-rays, a wavelength of light most people have likely encountered in

---

* Though not too short to activate similar cells in the eyes of most bird and insect species, which can see UV light as easily as we can see red, green and blue.

less-than-ideal circumstances. My first X-ray was at age eight, after an attempt to fly off a climbing frame like Superman resulted in a suspected fracture. While optical light bounces off your body (and clothes, upholding common decency) and UV light is absorbed quickly by the outer layers of skin, X-rays travel through your body – although only through your soft tissues, as they are almost totally absorbed by your bones. When you have an X-ray, the detector behind your body registers the light that passes through, thereby imaging your bones in silhouette.

Beyond X-rays lie the gamma rays, light so energetic that you have hopefully never been exposed to it. Gamma rays can be devastating to life because the light has high enough energy to cause severe mutations in DNA and kill living cells. Even this destructive property we have harnessed to our advantage, though. 'Gamma knife' surgery uses highly focused beams of gamma light to kill and shrink cancerous tumours deep within the body without ever needing to use a scalpel. Comic books might have you believe that a dose of gamma radiation will turn a curious scientist into the Incredible Hulk. Expose yourself to a source of gamma rays without medical supervision, however, and the damage to your DNA will mean you will likely not survive the week.

Let us move our gaze back towards the familiar optical wavelengths. Passing from UV to violet, back through the rainbow, past the reds and then into the infrared, we can relax a bit. The energies at this end of the spectrum are too low to ionise atoms. The best the light at this end of the spectrum can do is pass on its energy and get atoms a bit excited, so that they move and vibrate. On a hot summer's day, it is the infrared light that we feel on our skin, as it transfers its energy into our skin cells, gently warming us. While optical light interacts with our eyes and perhaps stimulates us to don some stylish sunglasses, we cannot *feel* it – just as we cannot feel the strength of UV light, or an X-ray.

Next in line after infrared are *microwaves*, the wavelength humanity has harnessed for the noblest of causes: reheating last night's leftovers. Microwaves are of just the right wavelength to induce frenzied movement in water molecules, producing so much heat as a by-product that food is quickly heated (and dehydrated). While microwaves are too low-energy to qualify as ionising radiation, you might notice microwave ovens are shielded in a way that your lightbulb, mobile phone and radio are not. Without this precaution, humans could potentially find some of their cells cooked along with the food, since our bodies contain so much water.

Finally, we finish our tour of the electromagnetic spectrum in the radio wavelengths – the longest of the lot. They are low-energy enough that they can't ionise your cells, or get water molecules excited; they don't even give you a warm fuzzy feeling like infrared light. They are, comparatively, the most inert grouping of the spectrum: they are not immediately dangerous to human life, or able to reheat a meal. But their subtlety is exactly what makes them extraordinary.

It can feel a little tribal to declare a strong allegiance to radio light, especially when the spectrum is continuous and no real separation exists between our categories. The range of wavelengths and energies results in new dimensions to our sensory world, adding so much depth and richness. But there really is something very special about the longest wavelengths. The property that separates them from their more energetic siblings is their ability to travel long distances relatively undeterred, through space, the atmosphere, clouds, rain, buildings and indeed us. This makes them invaluable in daily life as a communications tool, and in astronomy as the ultimate long-distance way of illuminating the Universe.

The degree to which light can move through the turbulent atmosphere and into our telescopes depends on its wavelength. For the most part, the protective layers of oxygen, nitrogen, carbon dioxide and water vapour absorb shorter wavelengths like UV light, X-rays and gamma rays. Crucially, this shields us from ionising radiation, which would otherwise pose a serious threat to life on Earth.

While the shorter wavelengths struggle through the atmosphere like celebrities pushing through paparazzi, optical light moves with far less resistance, like a member of the entourage. It gets jostled, certainly, but it makes it through, hair tousled, only slightly knocked off course. And a good thing too. Without our atmosphere's partial transparency to optical and infrared light, Earth would be plunged into an eternal, freezing night, without even the faintest twinkle of starlight. Even if humans were able to rely entirely on manufactured sources of light and heat, we would not last long. Photosynthesis, the process whereby plants use sunlight to convert water and carbon dioxide into oxygen and glucose would halt, destroying our source of food and oxygen in one fell swoop.

Fortunately for our survival, enough optical and infrared light is able to filter down to Earth to keep these processes going and keep us warm. However, the way in which it is scattered as it travels through the atmosphere presents a challenge for astronomers. Even the casual stargazer can see the effect of this distortion – it is why the stars appear to twinkle. As the atoms move about in the atmosphere, they divert the path of incoming light in random directions, ever so slightly. This is similar to looking at a dropped coin at the bottom of a swimming pool: its position appears to change constantly, despite it lying quite still. Professional optical astronomy requires observatories at high altitudes, coupled with software to correct for the remaining distortions. And for the faintest, most distant optical targets, only space-based telescopes such as the James Webb Space Telescope (JWST) have a clear view.

Light from a faraway galaxy might consist of a rich mix of wavelengths that have travelled in tight formation for millions of light-years or more. But when it reaches Earth's atmosphere, much of this information is lost. X-rays, gamma rays and UV light are completely absorbed, while optical light is badly distorted.

Radio light is alone in the spectrum for its skill in moving through the scrum with anonymity, with any atoms in the way barely giving it a second glance. Radio waves remain in tight formation as they travel almost unimpeded through the atmosphere . . . barely noticing any clouds that might be in the way. This indifference to the weather is of huge benefit to radio astronomers, saving them time and sanity. Trust me when I say you do not want to be anywhere near an astronomer who has travelled all the way to Hawai'i, sat for five days in a high-altitude control room, far from the beach, only to return without a kilobyte of data thanks to the cloudy weather. In addition, observing in the radio wavelengths can significantly increase the time you have available for observations. While optical astronomers have to wait for the Sun to set in order to see any other stars, radio astronomers can observe through the day and through the night.*

The optical astronomer is forever searching for sightlines clear of anything that might stop an optical photon in its tracks after a journey of light-years across space: rain, buildings, telegraph poles, satellites. To a radio photon, however, that cloud, that house, that human, are almost as irrelevant as the atmosphere that felled its counterparts in other wavelengths. The radio photons continue their long journeys, bringing with them messages from other planets, other galaxies and even, as we will see, other eras – messages we can receive and decode right here on Earth, day or night.

* Having said that, I have found radio astronomy tends to select for people like me who never were able to cope with staying up late, even to see a lunar eclipse. One can generally gauge whether an astronomer observes in the optical or radio by counting the number of coffees they need to get through the day.

How to Build a Radio Telescope

My fascination with radio began in my early years. I grew up in a three-bedroom semi-detached house in the suburbs of a non-descript town in England. I was an only child, but the house was far from quiet, thanks to the sound of the radio in every room. They were everywhere – in the kitchen, bathroom, bedrooms, even on the landing and in the hallway, and they were always on. You could easily walk around the house and find yourself listening to the same news report or farm-based drama uninterrupted. In this way, we were pioneers of 'seamless' entertainment. The default station was usually BBC Radio 4, but I enjoyed slowly turning the dial on our older radios, listening as each station emerged from the static, music of every kind sharpening into focus before fading away as I surfed the radio waves. I loved that the movement of the dial itself had a resistance to it, you couldn't just press a button to skip to the station you wanted. Instead, you had to sample the stations, just for a moment. And sometimes, along the way, you would make an unexpected discovery: a burst of music so new and surprising it would seize your soul and freeze your hand mid-turn.

When I was six or seven years old, the antennas projecting from the radios were just as much of a marvel to me. I'd experiment by angling them in different directions, and extend and retract them over and over.* My curiosity once led me to pull a telescopic antenna out too far, and, expecting to find some miraculous gadgetry inside, I was perplexed to discover an ordinary hollow metal tube in my hand. What's more, the radio fell immediately silent. How was it that this appeared to be the most indispensable component?

The variety of shapes was puzzling too: we had portable black

* Did I mention I was an only child? We have to find our own entertainment.

units powered by batteries and bulky, colourful monstrosities, with a wire loop for an antenna, plugged into the wall. We even owned a couple of hand-cranked versions, which would deliver snippets of broadcasts with no obvious source of power to a child yet to learn about energy transfer. My understanding was eventually pushed to its limits. Maybe I never asked my dad how they worked, or maybe he explained a little. But we had a rule in my house where you looked things up or broke things down to work it out for yourself, because it 'helped you learn'.

In the garden, within the sanctum of Dad's Shed, the answers were all there, amongst open radio cases and screwdrivers, cogs from half-made watches littering the workbenches, and bent antennas under repair in vices. Maybe with more time I would have dug into the mechanics of radio technology that early in life. I certainly had all the tools to do so, but it was not to be. I was distracted instead by a new, sure-to-be-fleeting fad called the internet. Radios were to remain filed in my mental folder of Mysteries for a long time, along with Santa's methods of surveillance and how that weird plastic fish from a Christmas cracker could tell my future.

At university, I developed a fierce passion for cosmology and set my sights on gathering light from the furthest reaches of space. I vividly remember the first day of my PhD. I was nervous, but ready to start my research career. It would be my challenge to unearth the first stars and black holes to exist in our Universe. How exciting! 'But how?' I asked on that first day. Would it be with the hulking great Hubble Space Telescope? The Very Large Telescope in the Atacama Desert of Chile? My face must have been a picture when I was told that I would spend my career not with these giants of optical astronomy, but fiddling with a radio telescope in the wet, muddy flatlands of the Netherlands. I won't lie, I questioned my choice of PhD programme a little as I finished my first day and forlornly looked up the Dutch for 'disappointment'.

Over fifteen years later I am writing a book on radio astronomy, so what happened? In short, I fell in love. What I had dismissed as an unglamorous relation of optical astronomy revealed itself to be nothing short of the stuff of science-fiction. Quietly plodding along in the background for fifty years, finding its feet on the ground as its optical cousins launched into the sky, I was let into the secret that it would be the radio telescopes that heralded the new era of astrophysics and cosmology. Indeed, to study the Universe at its earliest times, only a radio telescope would do. On that first day of my PhD, though, even with four years of education in physics and astronomy under my belt, I had no idea how one worked.

Radios are, in effect, translation devices. A microphone converts the sound waves of the newsreader's voice into an electrical signal, which is sent through an antenna causing the emission of electromagnetic waves in the radio wavelengths. We convert from sound to light, because these two forms of energy behave very differently. Sound waves need a medium like air or water to travel, passing energy on in a chain of collisions from speaker to listener. Energy is quickly lost along the way, especially when travelling through a dense medium such as walls: even the loudest neighbour will be inaudible to someone sat in the next building. Electromagnetic waves, however, need no such medium. They carry energy within themselves, encoded in photons, allowing them to travel efficiently through the emptiness of space. These waves cross towns, countries and sometimes continents, inducing an electrical signal in any antenna that has been tuned to receive them. Finally, a speaker translates that electrical signal back into sound waves, which our ears have evolved to hear.

A radio telescope operates on similar principles, using light to carry information, but without the need for that first step of sound-to-light translation. Unlike our radio presenter, the Universe already speaks the language of light. Yet for humans wanting to understand these cosmic radio waves, some translation at the other end is still required. We can, and do, hook up a speaker and convert the radio light into sound, hearing passing meteors as pleasing 'pings', for example. Mostly, though, we assign the radio wavelengths equivalent visible wavelengths, producing radio images translated to an optical colour scheme that our eyes can see. This is like a 'paint by numbers', where you are given a key to translate the coded canvas: 3 equals red, 4 equals blue and so on. It doesn't matter whether we choose 2 m to be equivalent to 400 nm (violet) or 700 nm (red), or any colour in between, as long as we are consistent.

Without radio communication, tasks like connecting to Wi-Fi to upload that Instagram post, looking up a map to the nearest pub or ensuring precise GPS targets for a nuclear submarine would be far trickier. You might not realise you are carrying a radio in your pocket, because the long, extending antennas have been miniaturised into tiny components, tucked deep within your mobile phone. But whenever you make a call or browse the internet, these antennas are picking up radio waves, and not just from terrestrial broadcasting towers and satellites. Beneath all those memes and messages are stowaway signals from the cosmos. Congratulations, you are a radio astronomer, and you didn't even know it!

A purpose-built radio telescope differs only in scale and shape, not in spirit, from any of the antennas you may have come across in your life. Because the cosmic signals are so much quieter than the human-made radio signals that course around us, astronomers place their antennas in remote locations, away from the radio noise that accompanies human activity. Even then, the light is so faint we need to gather as much of it as possible, so we often place the antenna at the centre of a metal dish. This serves the purpose of capturing incoming signals from a wider area, reflecting them onto the antenna like a reverse megaphone.

As a rule of thumb, the bigger the dish, the fainter (and so the further away) the signal you will be able to detect. The radio telescope local to me, the Lovell Telescope at Jodrell Bank, consists of a 76-m-diameter metal dish positioned on a circular railway track and a tilting mount. It dominates the landscape, a giant, 1,500-t bowl held aloft at a height of 90 m by Eiffel Tower-like scaffolding. Ladders criss-cross the support towers, leading to a disused (and frankly dangerous-looking) control cabin, and at the centre of that dish, there is a simple metal antenna extending outwards over 20 m. The scale is quite something to behold, even before you find out that it is listening to the heartbeats of spinning stars.

I adore the diversity in radio telescopes, a diversity bred from the sheer range of wavelengths covered by the radio spectrum. The optical wavelengths cover a tiny range from 400–700 nm. In terms of the physics behind capturing light, nothing much changes within that range. But when it comes to radio telescopes, one might be attempting to capture radiation with wavelengths tens of kilometres long, from the dawn of time, or mere centimetres long, from the death throes of a star. An antenna sized and shaped to be sensitive to a photon of wavelength 1 m will be almost entirely insensitive to a photon of wavelength 1 km.

Antennas are bent into all kinds of different shapes to pick up more specific wavelengths: there are long lines of wire to detect storms on Jupiter, and sheets of metal slotted together in a bow-tie shape to map the Milky Way. There is a radio telescope that looks for all the world like a simple metal table, shaped to be sensitive to the photons from the first stars. There is diversity in the dishes too. There are dishes shaped with spherical curvature, ideal for surveying large sections of sky, and dishes with parabolic curvature better suited for pinpointing specific stars and galaxies. Then, there are long pyramidal horn dishes, flaring open like funnels: their shape shields the detector inside from unwanted background signals, in order to reveal the faintest of cosmological sources.

We can even join these individual systems together, such that we have hundreds of thousands of antennas connected over kilometres of ground, creating a colossal telescope. While each antenna observes the sky independently, we can synchronise their signals to construct a far more detailed and expansive image than any individual element could achieve.

Radio astronomers opt for ground-based telescopes not because it is hard to go to space, but because it is easy to stay on Earth. Since the atmosphere is less of an obstacle to radio waves, we simply do not need to venture off-planet, and all of our funds and ingenuity can be poured instead into powerful antennas and state-of-the-art computing.

One might be tempted to equate radio astronomy's lower budget with it being somehow less advanced. But this would be short-sighted. Ground-based radio telescopes enable us to explore epochs and processes that surpass the reach of any rocket, probe or space telescope.

Not only that, all those billion-dollar spacecraft would be little more than decorative baubles in the sky without ground-based radio telescopes. All the data, all the instructions vital for pointing the telescope this way or that, must be communicated through Earth's atmosphere using radio waves. In principle, we could talk to our spacecraft using optical lasers, but that demands a clear sky and an accuracy so fine it is like threading a needle from millions of kilometres away. Radio, by contrast, is robust and reliable, and so will always be needed for critical space communication. Without it, the stories of the Moon landings, the space station, JWST and every other step into the unknown would have been starkly different, and mostly impossible.

Radio waves can pierce the thick atmosphere of Venus, uncover underground lakes beneath the Martian surface and peer through the bright heart of the Milky Way to the spiral arms beyond. Radio photons are survivors. Despite the turbulence and dust that have filled the Universe over its 14-billion-year history, radio light from the Era of the First Stars has roamed across cosmic time and space, reaching Earth still whispering tales of newborn stars and baby black holes.

A Day in the Life of a Radio Telescope

I have stood beside plenty of radio telescopes and never been able to make more than an educated guess as to what they were observing. That's if I tried to guess at all – I am often too giddy just from seeing the machinery. It is the same feeling I get when standing in front of an Apollo-era space capsule in a science museum: my mind struggles to connect something so tangible, so rusted and compact, with events of such astonishing scale. Visiting radio telescopes just to gawp at them has become something of a hobby, and I have photographic proof in an ever-expanding album labelled 'Emma with Radio Telescopes'. I stand awestruck next to the grandeur of the 305-m dish in Arecibo, Puerto Rico; I grin with delight as I hold my arms outstretched to give scale to the bow-tie-shaped antennas barely grazing my ankles in the Netherlands; I peer into the sky along with the dish standing next to me in California. These telescopes, all sitting seemingly inert, could have been doing any number of things as I bounced around at their bases. They are astonishingly versatile machines.

For example, a day in the life of a radio telescope might unfold as follows:

7 a.m.–12 p.m.: Search for signals from the Universe's first stars.

12 p.m.–3 p.m.: Monitor the Sun for sunspots and potential solar flares.

3 p.m.–6 p.m.: Employ radar to monitor the path of asteroids that could pose a threat to Earth.

6 p.m.–9 p.m.: Examine exoplanets for any hints of extraterrestrial communication.

9 p.m.–12 a.m.: Indirectly detect gravitational waves by observing the movement of radio stars.

12 a.m.–3 a.m.: Map gas clouds to outline a newly discovered spiral arm of the Milky Way.

3 a.m.–6 a.m.: Use radar for high-precision mapping of future lunar landing sites.

6 a.m.–7 a.m.: Revisit exoplanets for alien signals, just to double-check.

Notice there is no down-time during the day, unlike for optical telescopes which are mostly idle while the Sun is up. In fact, the daytime is when radio astronomers are in charge of protecting the planet from all sorts of nastiness, like solar flares that could take down our communications satellites and asteroids that could take down . . . well, all life on Earth.

Optical astronomers have to contend with gruelling all-nighters, altitude sickness and entire research trips being written off by drizzle. Their observation schedules, booked months or even a year in advance, offer no second chances. Sometimes you must go home exhausted, empty hard drives in hand, and apply again next year. Radio astronomers, in comparison, have double the time available in the schedule, reduced impact of weather on observations and more opportunity to take observations remotely or automatically. I have controlled a 100-m radio dish in West Virginia called the Green Bank Telescope from my bedroom in London. I was in my pyjamas eating popcorn as it tracked a radio galaxy across the sky. Perfection.

Radio telescopes can act as both 'receivers' and 'transmitters', sometimes switching between the two modes several times a day, or even within minutes. As with a telephone, we use the same hardware for both 'speaking' and 'listening'. For a lot of this book, we will set our radio devices to listening mode and gather up the faint photons emitted from a faraway galaxy or perhaps from within our own Milky Way. These radio photons will have travelled mostly undisturbed for anything from minutes to billions of years, depending on their origin. Eventually, they strike a dish and are reflected onto the waiting antenna. When the oscillating radio waves cause the electrons within our listening antenna to move, creating a current that we then convert into an electrical signal, we can begin the business of decoding the message.

But before we begin listening to messages, we are going to strike up a conversation with our neighbours. Setting up our radio equipment to transmit a signal, we are going to venture out into our Solar System, without leaving Earth: dishes and antennas will serve as our rocket ship. When we flip the switch, sending a current through the antenna, radio waves will be generated from that strange swaying of electrons. They will course through the air, through the atmosphere, not stopping to shiver as they enter the cold depths of space, until they meet something solid enough to act as a mirror, scattering a fraction of the signal back to Earth. What reflects one radio wavelength might be transparent to another, and by tuning our telescope we can choose which partners in conversation to approach.

As we get ready to launch our radio waves, remember that there won't be a sound, and we won't see a beam of light burst forth. There will be no clouds of dust or flickers of flame from a rising rocket. Our antenna just sits there, seemingly doing nothing at all. But with that flip of the switch, it begins to send out waves of radio light. Almost immediately, these waves crash upon the surface of our next-door neighbour, the Moon – the first stop on our tour of the Universe.

Bouncing off the cold, cratered lunar crust, that first pulse of radio light begins its journey home, carrying with it priceless information. To our eye, the telescope hasn't stirred. But now it has stopped transmitting, and instead it listens, trying to gather up the faintest echo of light from a Moon so near, yet still so full of mystery.

OUR SOLAR SYSTEM

1

THE MOON

Neil Armstrong: Magnificent sight out here.
Buzz Aldrin: Magnificent desolation.

Apollo 11 Mission, 20 July 1969

The days prior to 20 July 1969 were gruelling for Pat, Joan and Janet. They were prisoners in their own homes, hounded by the public and press alike if they dared open their front doors. They distracted themselves in any way they could, cleaning their houses and chain-smoking cigarettes. Perhaps they pushed away their anxious thoughts by turning on the record player, each trying not to count the seconds since the launch that had severed connection with their husbands so completely. Mike, Buzz and Neil were impossibly far away, and only getting further, on what everyone could only hope would be a return mission.

On 12 September 1962, John F. Kennedy, before an assembly of 40,000 in Texas, had announced the most audacious and ambitious technological endeavour in history. 'We choose to go to the Moon . . . We choose to go to the Moon in this decade and do the other things, not because they are easy, but because they are hard.' Kennedy had been right that travelling to the Moon would be neither easy nor cheap. Project Mercury, the first step that carried Americans into Earth's orbit, cost the equivalent of about $2 billion today. The Apollo programme that followed would demand far more, eventually running into the hundreds of billions of dollars. There was a human cost too; three astronauts tragically lost their lives while testing the *Apollo 1* capsule on the launchpad.

Despite setbacks and tragedy, NASA pressed forward at astonishing speed, each mission a chance to press closer to the Moon, to build knowledge, confidence and suspense. The crew of *Apollo 10* had flown within 9 miles of the lunar surface, and now, less than a decade after JFK's ambitious speech, NASA brought it all together and launched *Apollo 11*, with the intention of attempting the landing for real. Humanity now stood on the threshold of stepping onto another world and could only watch and wait. For the crew, by contrast, the three-day journey was full of activity. There were numerous tests and checks to carry out, and barely time to look out of the window. Every second was spoken for, every distraction an indulgence to be justified. Cigarettes were certainly off-limits – understandably, considering the men were sitting atop tanks of rocket fuel – but so were most other forms of entertainment.

It takes an enormous amount of thrust to escape the pull of Earth's gravity, and the heavier the rocket, the more fuel required. Adding just a kilogramme to the mission cargo would add thousands of dollars to the budget. Vinyl albums were out of the question, let alone the heavy devices required to play them.

Instead, they were treated to the newest of technologies, not available on the mass market – a compact cassette player.

In choosing the music, the astronauts shunned the harder-edged rock of the late 1960s, opting for classical pieces or folk tunes instead. Armstrong, who arguably had the more complex taste in music, chose to bring *Music Out of the Moon: Music Unusual Featuring the Theremin – Themes by Harry Revel.* Even the lightweight cassette players were only allowed onboard because they doubled as recording devices for notes and observations – the musicians of Earth gradually giving way to the first Moon voyagers.

Despite the soothing tones of classical music, the ever-present danger must have found a way to penetrate their minds. Journalists coupled photographs of stoic wives with morbid features describing how, if the trajectories were even just slightly off, the lunar module would either slam into the surface or drift into deep space and become a coffin. There was no shortage of terrible scenarios with which to torture themselves, for both the astronauts and their families.

For Pat, Joan and Janet, the Moon had suddenly revealed itself to be a potential enemy, a lethal desert island. This was a sudden transformation to stomach. Once vital for navigation and timekeeping, the Moon had been made obsolete in daily life by satellites and atomic clocks. It seemed to many that the Moon had few mysteries left worth solving. Yet for NASA, there remained a puzzle of the highest stakes: could three fragile human beings land on its surface – and survive?

Thinking of the lunar surface as hostile was anathema to poets and romantics, perhaps, but for the astronauts, it was an explorer's instinct, necessary for survival. Luckily for them, Neil Armstrong, Buzz Aldrin and Michael Collins were not the first envoys humanity had successfully landed on the Moon. Before Apollo, radio astronomers had already reached across space with their instruments, probing the lunar surface. These early observations helped convince NASA, and the president, that a moon landing was possible, and with that came a heavy feeling of responsibility. When Armstrong announced that they had landed safely, then, it was not just the astronauts' wives who breathed a sigh of relief. Before Neil Armstrong took his one small step, radio astronomers had already made a giant leap for science.

The Moon is an easy enough subject to observe. One does not have to be an astronomer, amateur or professional. It is easily visible with the naked eye, outshining the brightest star in the night sky, Sirius, by a factor of 40,000. Unlike a star, though, the Moon does not produce its own optical light. Like planets and other moons, it is far too small for the weight at its core to ignite fusion, the internal mechanism that causes stars to shine. Instead, the Moon reflects sunlight onto Earth.

One half of the Moon's surface is always lit by the Sun, just as one half of Earth is always in daylight. But how much of that lit half we see from Earth changes as the Moon orbits us. This shifting view gives rise to the familiar phases: Waxing Crescent, First Quarter, Waxing Gibbous, Full Moon, Waning Gibbous, Last Quarter, Waning Crescent and New Moon. It is during this last lunar phase that we see just how much light is produced by the Moon itself – none. In the New Moon phase, the Moon is positioned between the Sun and Earth, such that the side facing Earth is not illuminated at all: it is right there in the sky, but optically invisible.* In the radio wavelengths, however, the situation is quite different.

The Moon produces its own radio light, so the radio Moon

* The exception to this is during a solar eclipse. The Moon's plane of orbit is tilted at five degrees, which means that the Moon usually doesn't pass exactly in front of the Sun, and so remains invisible. But, occasionally, the position of Earth and the Moon line up just so, such that some people on Earth get the treat of seeing the Moon block the Sun as it moves across the sky.

always looks like a Full Moon. We can observe it at any time of the month, and even during the daytime, as long as it is above the horizon. But while radio moonlight is not a reflection of sunlight, there is still a connection between the two. After all, the energy to create radio waves has to come from somewhere.

Whenever matter is heated, the particles making up that matter move about more than usual. This thermal motion produces light across the electromagnetic spectrum, which we call *thermal emission.* Thermal emission tends to emit most strongly in longer wavelengths, which is why the police use infrared cameras to spot human bodies as they flee from the law in the dead of night.* Humans, like moons, are not natural light sources. Stick us in a dark room and we do not glow. Not in the optical, anyway. In the infrared, though, humans glow like fireflies because we are naturally warm. When it comes to the Moon, that warmth comes from the Sun.

Although the Moon reflects enough sunlight during a full Moon to cast shadows on Earth, that is only 10 per cent of the light that hits the lunar surface. The other 90 per cent is absorbed. All the energy from that solar radiation heats the lunar surface, producing thermal emission that peaks in the infrared and radio wavelengths.

The properties of the Moon's thermal emission depend on the lunar material. The Sun acts like a musical conductor, stimulating the orchestra of molecules on the lunar surface. Within the music, we can clearly pick out the instruments in that orchestra: the wavelengths and temperatures of the thermal emission reveal the composition of the lunar surface, allowing us to discern between solid rock or quicksand that would envelop any lunar lander in one greedy, dusty gulp.

In 1946, over twenty years before man first stepped on the

* Not even stone-cold criminals can hide from the law of thermal emission.

lunar surface, physicists in the US detected thermal emission of radio waves from the Moon for the first time, and used its spectrum to estimate a temperature for the lunar surface: −50 °C.[1] In thermal emission, we consistently observe a full disc. However, its brightness depends on the temperature of the Moon's surface: the more it is illuminated by the Sun, the higher its temperature. Having observed this with infrared radiation, the radio astronomers were never expecting the thermal emission in the radio wavelengths to be constant.

A group of physicists in Sydney, Australia, began to take measurements over longer periods of time and found that the intensity of the radio light from the Moon varied along with the lunar phases, as expected. Sunlit areas can reach about 120 °C, while temperatures drop to −170 °C in the shadows. But what was strange was that the radio temperature appeared to have a three-day lag, reaching its peak three days *after* the Full Moon. Similarly, the radio emission was at its lowest when a crescent was already back in the sky, exactly three days after the New Moon. The physicists puzzled over this for months. It was akin to turning off an oven and seeing the temperature keep climbing for several hours.

We knew very little about the Moon back in the 1940s, and our guesses as to what it might be made of ranged from volcanic rock to green cheese. Astronomers realised they could explain both the infrared and radio observations if the lunar surface was made of two components.[2] The surface layer, called *regolith*, would need to consist of a material that was a very poor thermal conductor, a granular mix of materials such as basalt or pumice. Its loose structure would act as a layer of insulation, reflecting infrared in tandem with the lunar phase.

The astronomers then used radio waves to effectively dig down into the regolith. The depth we reach depends on the wavelength of the radio light – so at a wavelength of 1 cm we're sensing the light from roughly 1 cm below the surface. Just like sunlight, the spectrum of radio moonlight allows us to estimate temperature, and so we can build up a temperature gradient as we change wavelength and dig. The thicker the regolith, the better it insulates, meaning it takes longer for heat to travel down and radiate back out as thermal emission. The scientists had initially been surprised by this power of radio waves to act as a remote excavation tool. But, once accustomed to the idea, they realised that the three-day lag was short enough to suggest a thin layer of grainy regolith as opposed to a fine dust, a surface that would support the weight of a lander.

The existence of the deep, dusty sandpits that Pat, Joan and Janet might have feared would swallow their husbands was quickly ruled out. NASA could be confident that man's first step on the Moon would be on solid ground, thanks in part to observations by radio astronomers a whole twenty years before the launch of *Apollo 11*.

The launch of *Apollo 11* was about politicians using a futuristic space race as a proxy for war, following the embarrassment of so many space-based firsts achieved by Soviet Russia. In contrast, future *Apollo* missions were scientific endeavours. They deployed seismometers to measure moonquakes, and captured the solar wind in aluminium foil. One memorable experiment had an astronaut drop a feather and a hammer to demonstrate that, with the lack of air resistance in the vacuum of space, the objects hit the lunar surface at the same time (they really do, you can watch the video). Mostly, though, scientists wanted to know how the Moon had formed, and what relation it had to our own planet's history.

The *Apollo* missions all landed in different areas on the Moon's surface, in an effort to sample different geology. But the astronauts were severely restricted in how far they could travel. *Apollo 11*'s furthest excursion was 50 m and, even with the advent of the moon buggy, the crew members never ventured further than 10 km from their lunar landers. They had to be selective about what they brought home too as, due to weight considerations, NASA had a carry-on policy limit stricter than a budget airline. In all, the *Apollo* missions brought back just 382 kg (842 lbs) of moon rock, which is barely enough to fill two bathtubs, let alone share out amongst the world's scientists.

I am lucky enough to have held samples of moon dust in my hands, having applied to 'Borrow the Moon' from a UK research council, but only after signing comprehensive risk assessments.[3] And, to be clear, the assessments concern the risk people pose to the moon dust, not the other way around . . . we are far less valuable. In order to qualify for the scheme, I had to buy a new safe. And a new lock for the room that housed the new safe. Moon rock is one of the most precious commodities on Earth.

There is no question that holding a piece of moon rock, scooped up by an astronaut's own hands, is a special experience. These samples have revealed a great deal about the geology of those specific landing sites. But imagine trying to understand the diverse geology of the Grand Canyon, the Scottish Highlands and the Siberian tundra based on a few suitcases of rocks gathered from a handful of random locations on Earth. We are asking a lot from two bathtubs full of stones. Radio astronomers, in comparison, can carry back information from our lunar visits without reference to a weight limit.

It is thought that the Moon formed after a Mars-sized protoplanet, Theia, named after the Greek mother of the Moon, collided with Earth 4.5 billion years ago. The two bodies merged in a molten mess, while the debris that got trapped in orbit coalesced to form our Moon, a perfect mixture of its parent planets. While any volcanic activity has long since died down, this dead, desolate world has the stories of its past written upon it like hieroglyphs on the walls of a pyramid. We don't even have to do all the excavation ourselves. Impact craters from meteorites do half the job for us, while we have begun to use radio waves as a tool to dig even further.

With modern radio telescopes, we can detect the thermal radio emission of materials in individual basins and geographical features on the Moon, and discern whether these areas are composed of different kinds of rock. There is the basalt of the flat 'seas', a volcanic rock formed from the quick cooling of magma above the surface. These are the darker patches we see by eye on the Moon, such as the Sea of Tranquility, where *Apollo 11* landed. Then there is the whiter anorthosite of the lunar highlands, formed from when ancient magma cooled below the lunar surface, before rising like a foam to the surface and solidifying. The different surface compositions and thicknesses across the cratered surface result in different levels of thermal conductivity, changing the intensity of the radio emission and the length of the lag behind the lunar cycle.

Each crater is a record of an impact, whether from one of the golf balls chipped by Alan Shepard during *Apollo 14*, or an asteroid. We are all at the mercy of these space rocks, which roam the Solar System, colliding with planets and moons. Earth is no exception, though thanks to our atmosphere most asteroids burn up before striking the ground. Some rocks do still get through, but the evidence of these collisions is hard to find because of the dynamic nature of our planet. Under your feet, billions of years' worth of impacts have been, quite literally, wiped off the surface of Earth by tectonic and volcanic activity, weather erosion, the footfall of animals and the build-up of human civilisation. Even the 150-km-diameter crater associated with the asteroid that killed the dinosaurs remained hidden until geological investigations in the 1970s.

This constant terrestrial resurfacing means that 80 per cent of the Earth's surface is less than 200 million years old. By comparison, 99 per cent of the lunar surface has not changed for more than 3 *billion* years. The lunar surface is a time capsule. The Moon has no wind to weather its rocks, no vegetation to smooth its surface, nor water to fill its craters. Almost all changes to the Moon's surface are due to only two things: meteorites and man.* Armstrong's first footprint will, if subjected just to the daily sprinkling of micrometeorites, or 'space sand', still be there in one million years. On Earth, my footprints will last only until the next rainfall – so an hour at most, where I live.

The surface of Earth is a poor informant on the history of our planet, then. We have to dig, drill and scrape our way through the layers of geology. And even then, these layers have been subject to the pressures of time, as more dead vegetation and animal remains gather above, compressing, heating and distorting that record of our past. Above us, far from that tectonic torture, the Moon is ready to speak, an artefact from Old Earth.

So far, we have discussed the *passive* method of radio astronomy – detecting the thermal emission created by a celestial object. Most astronomy is passive because in most cases the star, planet or moon is the source of radio light. We set up our devices and listen, like reporters holding their recorders up at a press conference, and there is no chance to ask questions at the end. We cannot poke or prod a planet in a test tube or heat up black holes above a Bunsen burner. We must make do with eavesdropping on the gossip from a galaxy, or the soundbites from a supernova. We have to solve the puzzle with no control over when or where the next piece might come from.

But there is another branch of radio astronomy which we can use in our immediate neighbourhood, including on the Moon,

* And, in this new space age, hopefully soon we can add 'woman' to that list.

an *active* method, where we launch our own radio waves and listen for the echoes. These radio photons enjoy a considerably quicker journey than the *Apollo* astronauts – 1.5 seconds – before they rebound from the lunar surface and return a mere 3 seconds after they left Earth. Barely enough time for the scientists to blink. Yet these photons are fundamentally changed by their journey. Their strength, timing and sometimes wavelength are forever altered by the alien world they have touched.

The method is called RADAR, an acronym for RAdio Detection And Ranging. It is as close as we get to experimentation in astronomy, where we, to some degree, control what we are asking: where the radio waves are directed, at what wavelength they are sent, for how long. The properties of the reflected waves describe the distance, position, shape and trajectory, even the texture and composition, of the object they strike – in this case the Moon.

Radar is so ubiquitous in daily life that the acronym is no longer regularly used, and it has earned its place in the English language as a common noun in its own right. When you are driving in your car and activate cruise control or collision notifications, you are using the cheap, portable radar antennas placed around the chassis. Every time you walk through an automatic door, it is because a radio antenna nearby has picked up the echoes produced as your body reflects incoming radio waves. Radar is everywhere because it is a cheap, effective way of measuring small or large distances between objects and mapping an area whatever the weather. It is no wonder that scientists began to consider turning our antennas away from Earth and up to the Moon.

Radar experiments aimed at the Moon were considered soon after radar became practical in the 1930s, but it took the pressures of war in the 1940s to provide the technology, expertise and funding to attempt it. In the trenches of the First World War, radio was essential for communication, while in the Second World War radar became vital for detecting incoming bombers and submarines. Radio engineers and physicists were hired in droves and trained in radar. Once released back into civilian life following the end of war, these newly anointed radar physicists had some time to consider how to apply this new knowledge. More than a few looked up and wondered how radar could further our knowledge of astronomy, and our closest neighbour in particular.

Two groups are credited with the first detections of the Moon using radar: one led by John H. DeWitt Jr in the US, and another by Zoltán Bay in Hungary. The US team operated during a post-war economic boom, while the Hungarian effort took place in a country devastated by war and occupation. Despite these contrasting circumstances, both teams achieved remarkable success in radar astronomy. It was not easy: by the mid-twentieth century, radio communication had become mainstream, so there were plenty of radio signals in the air. This had the effect of adding a constant interference to signals to and from the Moon. Imagine sitting at the back of a theatre and listening to an actor projecting their voice from centre-stage. It is easy enough when the theatre is empty, but should the orchestra begin to play, and the theatre begin to fill with the chatter of an undisciplined audience, the actor's voice is inaudible. Sitting in the audience, we do not have the option of tuning out all the hubbub . . . but a radio astronomer does.

Each unwanted radio signal occurs over a certain stretch of wavelengths and for a particular duration. For example, a telephone call may last twenty seconds and be broadcast on one specific frequency. The advantage of using radar is that we can choose both the wavelength and the duration of the signal we send, and because we know our target, we also know when the echo should arrive. We know what to expect, and when, so we can use computer software to filter out signals that don't match those expectations.

The defining characteristic of a 'moonbounce' is that there is the three-second delay between the transmission of the radio wave from our radio antenna and the return of the echo. In 1945, both groups of physicists demonstrated, within months of each other, that they had detected echoes with exactly this delay.[4] While we might consider Neil Armstrong's foot to be humanity's first contact with the Moon, I'm more inclined to think of a burst of radio waves as our first envoy.

When we illuminate a target such as the Moon with radio waves, it acts like a giant, dusty mirror ball, scattering some radio light in directions far from the reach of our receiving radio dish. Those radio waves will travel off into deep space to have other adventures, of which we will never hear. In addition, as the beam of radio waves travels, it will spread out, like the beam from a torch in a dark room. The radio signal will always return far fainter than when we sent it. To pick up such a faint signal, we require large dishes to gather as much light as possible, and sensitive antennas to tease out the signal we want from amongst the noise of our society.

Historically, in Apollo times, our receiving antennas were basic, and our ability to compensate for human noise with software non-existent, so we had to shout really, really loudly: we needed powerful transmitters. The first moonbounce was achieved with a transmitter of 50,000 W, and that barely produced an echo.

Now, over fifty years on from the Apollo era, the revolution in technology and computational power has meant we have developed the antenna-equivalent of noise-cancelling headphones, and the software to filter out almost everything except the delayed radio signal we are expecting. The Green Bank Telescope now uses a radar antenna with a power similar to your microwave oven. We can hear so well we don't have to shout at all.

Achieving a moonbounce with today's technology is so easy that we can do it for fun. In 2014, a Dutch astronomer, his fiancée and their wedding party took over the control room of the Westerbork Synthesis Radio Telescope in the Netherlands.[5] He broadcast his 'I do' to the Moon using radio waves, and three seconds later, the echo was received, serving as a confirmation of their union and a practical implementation of the popular phrase, 'I love you to the Moon and back.'

A single moonbounce tells us about the patch of lunar terrain it reflects from, so, ideally, we want to gather multiple echoes from across the Moon's highlands and craters. The ability to distinguish between nearby echoes – from two adjacent craters, for example – is known as *resolution*. In simple terms, it's a telescope's sharpness of vision. The Green Bank Telescope can now produce radar maps from Earth that compete with optical maps taken *from lunar orbit*.[6]

Radar and optical maps are not directly equivalent, though, because they sample different depths. We can take a photograph and an X-ray of the same hand, for example, and they will depict entirely different structures in our anatomy. Radio waves can penetrate to a deeper distance than optical light, as they suffer far less scattering and absorption, such that they reach depths of 10 m or more. Radar images of the Moon let us assess landing sites for returning Moon missions, as they reveal dangerous structures lurking below the surface dust, hidden from plain sight.

Radar allows us to plan for longer lunar adventures than enjoyed by the *Apollo* astronauts. It is all very well getting to the Moon and having a walk around, but what then? Any longer-term moonbase missions are going to require the base necessities to preserve life: water and shelter.

It has long been suspected that the Moon is riddled with networks of caves, ancient lava tubes left hollow from long-drained and hardened magma flows. In 2009, optical images were taken of possible skylights – entrances into these lunar caves. The limitations faced by astronaut explorers would make exploring each potential cave a risky endeavour. Radar has no such limitations. In 2024 radar data finally confirmed that one of these openings did indeed lead to a subsurface chamber.[7] The radio waves probed into the entrance, echoing in the chamber and back to the Lunar Reconnaissance Orbiter. Only 400 km from Tranquility Base, this is a place our new generation of space explorers can run to when life gets tough on the Moon. When we need shelter from the radiation of a solar storm or micrometeorites are heading our way. When we need a break from the wild, 300 °C temperature variations over a lunar month. When our supplies of water have dwindled and we need to look for somewhere cool, where the Sun has never reached and water ice might exist within the walls. For their long-term survival, it is vital astronauts travel to the Moon armed with maps created by an earlier pioneer – the radio photon.

I cannot write about the ways in which radio waves have enabled lunar exploration without mentioning one very practical aspect. Without radio waves, the story of the space missions would be one of silence. All communication with spacecraft, then or now, uses the medium of radio waves, thanks to their ability to cut through Earth's atmosphere. As Pat, Joan and Janet tuned in to updates from mission control, they were listening to the voices of their husbands carried across the vacuum of space and transferred and transformed by one of a series of designated 'uplink/downlink' radio telescopes spread around the globe. There were even antennas on ships roaming the seas so that, should the *Apollo* missions choose to place a call, there would always be someone to listen and relay the message between ground-based stations, until mission control could pick up.

Without that means of communication, enabled by radio

telescopes, we would have launched *Apollo 11* and then suffered an eight-day wait for news of success or failure. The astronauts would have been entirely alone, cut off from the expertise at NASA that led them through every error code and reassured them they were 'Go' for landing. Had they not returned, we might never have known why. The three wives would have been like the loved ones of the great explorers of the high seas, desperately scanning the horizon for the signs of a returning sail, or in this case the parachute, that would signal the return of the lunar capsule to Earth. Aldrin, Armstrong and Collins would have returned with notes, logbooks, cassette recordings of thoughts and reels of video tape, so there would have been plenty of media to consume, eventually. But the world would never have come to a standstill, and sat down as one to watch Neil Armstrong take his first step live on television (albeit with the 1.5-second delay). That moment of global unity would have never happened, the news instead spread across time zones and news channels, eight days after the launch.

Over fifty years later, the Moon might seem to have few secrets left. After all, it is a celestial body we are so familiar with that we have walked its surface, played golf on it, planted flags on it, surrounded it with satellites and brought pieces of it back to put in our museums.

The crewed space programme to the Moon looms so large in our history that it is sobering to realise the Apollo programme lasted only three and a half years. Six missions carried a total of twenty-four American men to lunar orbit between 1969 and 1972, and of those, twelve set foot on the surface. While the landing sites were spread across the nearside, the total surface area of the Moon that humans have walked is minuscule. The crew of *Apollo 11* walked on the surface of the Moon for only two and a half hours, during which Armstrong never ventured more than 60 m from the lunar lander.

I wonder how Armstrong and Aldrin felt, looking at the lunar horizon and knowing they had effectively been told by their elders at NASA to 'stay where we can see you'. Not only that, but less than 2 per cent of their mission time was spent on the lunar surface. This is equivalent to making the nine-hour flight from London to Orlando and then telling your kids that they have three minutes to explore Disney World. For Michael Collins, commander of the lunar orbiter, it was even worse. He had driven all that time only to be told to stay in the car.

The Moon landings, as inspiring and exciting as they may be in the public imagination, were far from a thorough geological and exploratory survey. As we return to the Moon in the years to come, we will be able to further that exploration, but at great financial expense and risk to human life. Astronauts, unlike photons, are fragile and require the flattest of plateaus and the softest of landings.

Radar, in comparison, allows mistakes. We just reconfigure, adjust our aim, and try again. We can quite happily launch legions of photons with the express aim of smashing them against a surface so hard they bounce right back to Earth, and if there are casualties in the photons lost to deep space then so be it. Human life is precious, and space travel is expensive. Radio astronomy, on the other hand, is so cheap we can afford to send beams of radio waves to hit every slope, every crater and every hill, ensuring that when we do send crews out there again, it will be to the most scientifically interesting, and the safest, lunar locations.

If we want to truly understand the Moon – to truly walk across its vast and varied landscape, to understand our history and plan for our future as a space-faring species – then our best mode of travel is not the space rocket, but the radio telescope.

2

THE SUN

Light is not only glorious and sacred, it is voracious, carnivorous, unsparing. It devours the whole world impartially, without distinction.

J. M. W. Turner

Our sky is dominated by two celestial giants of very different character: the Moon and the Sun. By night, the Moon, with its calm and familiar face, takes on a prominent role, gently illuminating the world beneath it. Yet to our eyes, it is just one of many performers on the nighttime stage. The Moon is like an understudy, basking in borrowed sunlight, but too faint to outshine the vast, glittering chorus of stars beyond.

In the daytime, within moments of taking to the stage, the Sun's blazing presence banishes the stars, planets and often even the Moon itself from sight. The Sun's light is so powerful that it consumes our sky, hiding the rest of the cosmos behind a curtain of brightness from sunrise to sunset.

The Sun is the star of the daytime stage and, indeed, a literal star. It is a ball of hot gas and plasma that is collapsing under its own weight, fusing the hydrogen atoms at its core to create helium, releasing energy. This fusion process has sustained it for billions of years, pushing back against its collapsing layers and flooding our planet with light and warmth as a happy by-product.

With surface temperatures of about 5,500 °C, the Sun is a small and cool star. It is not an award-winner by any measure. There are stars far more massive, far brighter, far hotter, and stars far cooler and far lighter. There are stars that are far older, and stars that are newborn. In terms of the casting in our Solar System, however, we have really lucked out. For a start, there's only one Sun in our sky, which is abnormal, as most stars are in binary systems. The dynamics of orbiting two Suns is not trivial, and creating stable conditions in which life can thrive is more challenging.

Our star is also at its most stable in terms of its growth, 4.5 billion years into a 10-billion-year lifetime. Very young stars are volatile and prone to irradiate anything remotely nearby, while very old stars end their lives by swelling and enveloping their planetary systems, or obliterating them in huge explosions. They insist on taking the theatre down with them. Our star, thankfully, is in its prime, mature and steadfast. It gives a startlingly reliable performance, producing the same amount of optical light day in and day out. While we fly around the globe seeking out warmer temperatures for beach holidays and colder for ski holidays, these temperature differences are caused only by our location on the surface of the Earth being tilted towards or away from the Sun. The Sun itself has no winter or summer seasons.

But this consistent stage persona is not all there is to our Sun, as the radio version of this play reveals. The Sun's radio light is much weaker than its optical light, so weak that it does not dominate the sky. The radio sky has an ensemble cast consisting of galaxies, supernovae remnants, planets, the Moon and, yes, amongst them all in the daytime, our Sun. Despite this, the radio sun is a diva. While in the optical the Sun's output varies by less than 1 per cent, in the radio it is highly volatile, varying by a factor of 100,000. It rages and screams, sometimes loud enough to overcome our own radio communications on Earth. These radio bursts come from a different part of the Sun, an outer layer known as the corona. The Sun, it turns out, is no one-dimensional character, but an individual with layers of personality.

We are used to seeing the 'surface' of the Sun, the *photosphere*. It is not the surface in any physically tangible way, but we define it so because, below the photosphere, the gas making up the Sun is so dense that the photons cannot travel freely. They get absorbed and emitted in all kinds of directions, such that it is like trying to see the floor of a jacuzzi when the bubbles are turned on. The photosphere is the point at which the density of the gas falls enough for light to break through.

Usually, the Sun is so brilliantly bright that the retinas in our eyes, and the optics used in a lot of sensitive telescopes, can be permanently damaged by looking too closely. But there are rare moments when the Moon joins the Sun on stage, eclipsing it, in the ultimate cosmic cross-over. During these solar eclipses, the Moon perfectly aligns itself between the Earth and the Sun, casting a shadow that briefly plunges part of the world into an eerie twilight. Birds stop singing, and the temperature drops. For the few minutes it takes for the Moon to move along in its orbit, the relentless glare of the photosphere is tamed, revealing a mane of sunlight with tendrils extending millions of kilometres into space: the corona. It is a haunting sight, a dynamic and violent outer atmosphere of plasma which produces weak optical light and chaotic radio signals.

For most of human history, solar eclipses were our only

opportunities to glimpse this mysterious outer edge of the Sun, and even now, they remain one of nature's most awe-inspiring spectacles. In the radio sky, we can see the corona all the time, not just during an eclipse, because the radio photosphere is not as overwhelmingly bright as the optical wavelengths. The radio emissions from the corona are invaluable to scientists seeking to unravel its mysteries. Chief among them is its inexplicably high temperature. The core of the Sun, where fusion rages, is around 15 million °C. From there, the temperature steadily decreases as we move towards the surface, cooling to about 5,500 °C at the photosphere. But then, as one moves outwards into the corona, the temperature suddenly soars to over 1 million °C. This is as odd as moving away from a campfire and finding the air gets hotter the further you go. Why this happens is one of the great unsolved mysteries of solar physics. There are theories involving tiny bursts of energy called nanoflares and charged particles vibrating to conduct energy like sound waves, but there is no definitive answer. Radio observations enable scientists to observe the corona directly, without waiting on an elusive solar eclipse, perhaps one day helping solve this puzzle.

Understanding the mysteries of the corona is not just an obscure scientific pursuit, however. While radio waves are of low enough energy that we do not need to be afraid of them hurting us directly, solar radio bursts can overwhelm our terrestrial communications and infrastructure. Our dependence on radio communication technology has increased with dazzling pace over the decades, now underpinning GPS, Wi-Fi and mobile phones. Imagine a day cut off from all of that, an hour even. Imagine what would happen to the traffic, the stock exchange, the hospitals. As a species that depends on radio communication in nearly every part of daily life, we've begun to awaken to the invisible and existential threat posed by our own star, a threat borne on a wavelength we long believed it didn't emit at all.

★

The first effort to study the Sun's radio output came almost immediately after the discovery of radio waves by Heinrich Hertz in 1888. Hertz had shown they could travel across an empty room, and so space, using simple metal wires to transmit and receive them.[1] Physicists then attempted to detect solar radio waves using rudimentary antennas, but they failed so consistently that they abandoned the effort after fifteen years.

Instead, the antenna was used for pleasure and business, enabling long-distance phone calls where gossip and stock tips might be exchanged. Then, it became a device for defence and protection. SOS systems were placed on all passenger ships, saving countless lives, including on the *Titanic*. During the Second World War, radar stations were erected along the coasts of England, sweeping the skies with radio light and listening for the first faint echoes that signalled an incoming enemy bomber. It also became a means of attack, as radar systems were miniaturised and placed in planes. Allied bombers were able to locate German U-boats in the darkness. This technology allowed them to quietly approach and destroy enemy submarines, tipping the balance of the war in the Allies' favour.

But then, over two days in 1942, long after the notion of a radio sun had been dismissed, there was a mysterious series of blackouts at radar stations across England.[2] Military command was shaken. Engineers were already working to stay ahead of enemy efforts to 'jam' their signals. This method saw enemy fighter planes broadcasting loud radio signals on the same wavelengths used by the Allied radar network, interfering with the radio light echoes and severely hindering detection capabilities. This kind of jamming was localised and brief, however. These new blackouts affected vast sections of the defensive radar network, and for two whole days panic spread among the highest ranks as the country lay defenceless, disarmed of its only shield. Yet no attack came, and radar capabilities eventually returned to normal without any intervention. Radar engineers were left baffled. How had the

Germans managed to disable almost an entire radar network, and why hadn't they then exploited the chaos?

It took a physics teacher to work it out.

James Stanley Hey had been teaching physics at a boys' grammar school in England until, at the outbreak of war, he joined the Army Operations Research Group. After a six-week course on radar, the army deemed him expert enough and put him in charge of a research group working on the radar defence network. When this network was mysteriously blinded, Hey began to consider all possible sources of interfering radio waves. Upon tracking which stations were affected, and at what time of day, it quickly became apparent that the direction of maximum interference seemed to trace the movement of the Sun. Lacking any astronomical expertise, he made a phone call to the Royal Observatory Greenwich. The astronomers there informed him that a monstrous sunspot had made its way across the Sun on those days. They probably also informed him of the prevailing assumption that the Sun didn't emit radio waves. Unlike solar astronomers, who were familiar with the failed attempts to detect solar radio waves in the late nineteenth century, Hey wasn't burdened by this bias. He was free to connect the dots and conclude that, for two days, the war had apparently been at the mercy of a sunspot.

Sunspots are cooler areas of plasma on the Sun's photosphere, visible in optical light as dark patches, and in radio as bright active regions. While the Sun's surface can occasionally appear clear for days, more often than not, solar telescopes reveal several sunspots, sometimes dozens. From day to day, these sunspots seem to move across the Sun's surface, though it is the whole Sun that rotates. On average, the Sun rotates once about every 27 days, though it spins faster at the equator and slower near the poles. Longer-lived sunspots disappear around one side of the Sun, reappearing on the opposite side roughly two weeks later.

The link between sunspots on the photosphere and radio bursts from the corona is not immediately obvious. It involves an invisible force: a magnetic field. The Sun, a seething mass of charged particles, generates a strong magnetic field with looping field lines. To visualise this, think of the classic school experiment where iron filings are sprinkled around a bar magnet. I remember that experiment fondly because it's so quick and deeply satisfying; a pattern instantly appears that wasn't there before. The iron filings all act as tiny magnets, repelling each other and settling into open-ended lines and closed loops that trace the invisible magnetic field.

In the iron filing experiment, the magnetic field is created by a permanent magnet. In celestial bodies like the Sun, magnetic fields are generated by the movement of charged particles, so the magnetic field lines are constantly shifting. While we cannot walk around the Sun with a compass, or sprinkle iron filings around it, we can trace its magnetic field by other means.

Sunspots are found where the magnetic field on the photosphere is strongest. The heat from fusion causes convection, as gas heated in the Sun's core rises before cooling at the surface and sinking back down. Where the magnetic field lines gather, however, they hold the surface plasma in place, creating patches where the hot gas cannot break through. These regions are therefore 1,000–2,000 °C cooler than their surroundings, appearing about a third darker.

When Hey looked at the Sun, he did so with the eyes of a newly qualified radar physicist, and he recognised the electromagnetic mechanism behind sunspots. During the war, he had been working with *magnetrons*, machines that generate radio waves by combining a stream of charged particles with a magnetic field. Hey had the expertise to realise that, while the sunspots are a tell-tale sign of the presence of a strong magnetic field, they do not directly cause radio blackouts like those experienced in

1942. Instead, the same magnetic fields that create sunspots on the photosphere also cause separate radio bursts further out, in the corona.

With the rotation of the Sun, its magnetic field is constantly fluctuating. When magnetic field lines gather in one place, they find themselves suddenly forced into a new configuration according to the changing magnetic field, releasing huge amounts of magnetic energy in a process called 'magnetic reconnection'. Imagine trying to push two bar magnets together with like poles facing (N–N or S–S); you can feel the stored energy in the magnetic field. If you let go, the magnets will spin on their own, realigning to match N–S, S–N, and releasing that stored energy.

During magnetic reconnection, this energy accelerates surrounding charged particles in the corona, which emit bursts of light as they move through the magnetic field – just like in a magnetron. Although this light is often invisible to the human eye, radio telescopes can detect these 'radio bursts', allowing us to trace patterns in the solar magnetic field as if the bursts were iron filings falling around a bar magnet.

Sunspots have diameters ranging from 10 to 100,000 km, so large that the magnetic fields causing them could, during a particularly violent reconnection event, release enough radiation to overwhelm a radar network. We hadn't noticed before 1942 because it was wartime that had accelerated the development of radar. Before then, radio equipment was crude enough that any incoming bursts were recorded only as mysterious 'radio fizzlies' that would only cause passing concern to an amateur radio enthusiast, as opposed to widespread panic in the military. In addition, the Sun happened to be unusually active during those wartime years. The first forays into radio astronomy during the previous decade had occurred during a quieter time.

Astronomers had long busied themselves by counting sunspots, as a trainspotter might note down carriage numbers. If one plots the data, a clear pattern emerges, called the 'sunspot cycle'. The number of sunspots goes up and down and up and down, in an eleven-year cycle. I am tempted to refer to myself as having been born during Sunspot Cycle 22, giving a pleasing vagueness to my age.

This eleven-year sunspot cycle traces the complex reversal of the Sun's magnetic field. The Sun's regular flips are driven by the interaction between the different rotation speeds between the equator and poles and the magnetic fields in its convection zone. This process, known as the solar dynamo, follows a predictable, periodic cycle as the magnetic lines are dragged, twisted and pulled to breaking point. Over the course of several years, the churning and rotating Sun entangles the magnetic field lines to such an extent that they begin to reorganise themselves globally into a new configuration. Eventually, the old magnetic field structure collapses, and a new one forms with reversed polarity. This reversal marks the peak of the sunspot cycle. During these times, the surface of the Sun is dotted with dozens of sunspots, and radio bursts buffet the Earth.

When Hey wrote up his findings, his supervisor remembered an obscure research paper by a radio engineer named Karl Jansky. Hey had never heard of him but was delighted to find this field of 'radio astronomy' existed. Incidentally, when Jansky first heard the hissing of the Milky Way's centre in his headphones in the 1930s, he initially thought it was the Sun before ruling it out. At the time, the Sun was in a quiet period, and its signal was fainter than that of the black hole. Had there been more sunspots, Jansky would almost certainly have measured solar radio bursts and concluded that the Sun emitted radio waves, over a decade before the radar blackouts.

Hey couldn't prevent the solar radio bursts from continuing, but he took some solace in knowing they weren't a deliberate attack on England. He spent the rest of the war developing methods to counteract the more usual jamming techniques used by enemy planes, but he never could quite forget this new adversary in the sky: the Sun.

Due to wartime censorship, Hey's research papers on solar radio flares didn't reach the wider community until 1945. When he finally presented his findings, many astronomers were sceptical and indignant.[3] Who was this man, this *teacher*, to detect solar radio waves after their academic predecessors had deemed the exercise futile? So convinced were they that solar radio waves either did not exist or could not penetrate Earth's atmosphere, they dismissed the evidence in front of them. Fortunately, in February 1946, a massive sunspot appeared on the Sun and, at the same time, intense bursts of radiation were detected by radar equipment. The timing could not be explained away as coincidental a second time, and Hey's results were vindicated.

After the war, Hey left teaching and become a pioneer of radio astronomy. Solar radio astronomy became a fashionable field; Hey's work inspired astronomers around the world, including Ruby Payne-Scott in Australia.[4] One of the few women in the field, Payne-Scott made Australia's first radio astronomy measurement, and pioneered techniques that could pinpoint the exact origin of radio bursts.[5] For this, she earned immense respect from her male peers, no small feat for the time. Suddenly, astronomers had more than just the timing of a radio burst to go on. They could now trace a burst back to its source and, sure enough, that source was a sunspot.

Our Sun has a temper. In the optical, it is the picture of calm continuity, but in radio wavelengths, it behaves like a prima donna. The reconfigurations in the Sun's magnetic field trigger bursts of radio light that vary wildly in intensity and duration. As Hey saw, these solar temper tantrums can wreak havoc on our radio communications.

But when the Sun is really in the mood to rage, it doesn't just send light our way. It also hurls super-heated plasma and even magnetic fields at Earth, in events called coronal mass ejections (CMEs). During a CME, billions of tonnes of charged particles are launched into space at millions of kilometres per hour. Though fast, the movement of this plasma is still much slower than the speed of light. Radio bursts arrive within minutes, and so can act as a warning siren for CMEs that may not reach Earth until days later. It is an alert we would do well to heed: CMEs can disrupt our electrical infrastructure, short-circuiting power grids and causing Earth's atmosphere to swell, altering satellites' orbits, causing them to lose altitude and even burn up in the atmosphere.[6] For any humans outside the protection of our atmosphere, say on the International Space Station, or standing on the Moon, the high-energy particles associated with a CME can damage human tissue, causing radiation poisoning and cancers.

Our planet is threatened. Our astronauts are threatened. Even our internet is threatened, so any advance warning of a CME is valuable. Having played the scaremonger, I should note that you have probably lived through several sunspot cycles, complete with radio bursts and CMEs, and never noticed thanks to scientists correcting satellite orbits for atmospheric effects. Yet the signs are there: news of Starlink satellites falling out of orbit and warnings about power grid failures and internet outages all serve to remind us about the true might of the Sun at its most fierce.[7]

While our reliance on radio and electricity makes these events ever more apparent, one can even look back though history to see the cyclical impact of solar storms. The Carrington Event of 1859 was the most powerful solar storm on record. During this event, spectacular auroras were visible as far south as the Caribbean. Telegraph systems, the cutting-edge communication technology of the time, caught fire, with some operators receiving electric shocks.[8]

Knowing that a CME is on its way can save millions, if not

billions, of dollars of damage to our satellite and electrical infrastructure, but unfortunately it is not the case that every radio burst acts as a clear warning. The radio Sun can be extremely active without any CME involved, and some CMEs produce only a subdued radio burst. The stakes are high enough to keep studying the patterns, however, in the hope it can add to a comprehensive warning system.

When there is a particularly large solar event, those in charge of power grids might reroute power, reduce the load and shut down vulnerable parts of the network to prevent damage. Astronomers, meanwhile, with the hard work of predicting the event behind them, can simply look to the north, in the hope of glimpsing a glorious side-effect of solar tantrums: the aurora.

The Sun continuously discharges a stream of particles into space, known as the solar wind, which even on a quiet day travels at over 1 million km/h. When these charged particles collide with oxygen in Earth's atmosphere, they excite the atoms, causing them to emit a soft, green light. Earth's magnetic field funnels particles towards the poles, and so the effect is usually at its most vivid at very high latitudes. If you live far enough north, towards the top of Norway for example, then the sight of the aurora borealis, or Northern Lights, can become a familiar sight.

Even as far south as London, however, the aurora can still appear when the solar wind grows unusually strong, such as during a CME or at the peak of a sunspot cycle. At its best, the aurora sweeps across the sky in arcs and ribbons not only in green, but in vibrant reds and purples too, as more energetic collisions stir oxygen and nitrogen atoms at different altitudes in the atmosphere.

But the Sun is nothing if not temperamental. Aurorae are ethereal and mercurial. Even if you travel to the Arctic Circle during the peak of a sunspot cycle, there are no guarantees that your journey north will be rewarded. How do I know? Because I did just that, taking a job as an astronomer on a cruise ship sailing north, hoping for a dazzling display. I saw, at best, a faint haze, barely tinged green. Back home in England, meanwhile, my friends were treated to the best display in over a decade. Sometimes, you will get very lucky, and the solar activity will be so high that the colours are immediately recognisable and the lights positively dance across the sky. Not so for us. The days where the activity was high, so was the cloud cover.

It was my job to set people's expectations . . . and their camera settings. With a ten-second exposure, the light green mist seen with the naked eye appeared on the screen as a winding rainbow of light, streaking across a background tapestry of stars.

In my parting lecture to the passengers, I explained that our eyes are simply not built for gathering faint light; they quickly process incoming light into a series of isolated snapshot images. Cameras, though, can allow light to build up and intensify on the film or sensor. I empathised with my audience: it can feel like cheating to share these stunning, colourful images knowing you didn't see anything like it with your own eyes. But that is modern astronomy in a nutshell.

The iconic images from non-optical telescopes such as the infrared JWST or the Arecibo radio telescope are not awe-inspiring because they show us something far away. They are magnificent because they reveal what we cannot see for ourselves. The aurora is no different. Technology allows us to appreciate it in all its forms, whether it blazes red or flickers as the faintest wisp. Either way, it is awe-inspiring to see the Earth's atmosphere in action, shielding us from dangerous solar particles and dispersing their energy as light.

Since the radar blackouts of 1942, our world has only become more entangled with radio-based technology, and more desperate for ways to predict CMEs. Solar radio astronomy has never been about curiosity for curiosity's sake. It began as, and remains, a frontline defence.

The Sun is a life-giving star; we crave its presence and bask in its light and heat when it rises each day, right on time. But under and above that serene optical veneer is a seething surface of high-energy particles, trapped by twisting magnetic fields. At any moment it can flare outwards with devastating intensity, reaching tendrils outwards into the Solar System.

Today, the border we must protect is no longer drawn between countries, but between celestial bodies. After all, we do not need to look far from the Sun to be reminded about just how much we rely on our own magnetic shield. Our next destination lacks such defences. On the battlefield between Earth and the Sun, we find our scorched planetary brother: Mercury.

3

MERCURY

Reality is frequently inaccurate.

Douglas Adams, *The Restaurant at the End of the Universe*

Mercury has always been in a hurry, darting across the sky with a speed unmatched by any other planet. The Romans were reminded of their swift-footed messenger god and gave it the name it is known by today. Mercury, complete with wings on both his sandals and helmet, flitted between gods and mortals, delivering urgent news and admonishments from the heavens. The Romans were not the only ancient society to notice the planet's rapid movement. The Babylonians called the planet Nabu; the Egyptians knew it as Thoth: both were divine couriers reflecting Mercury's extraordinary speed.

Mythological considerations aside, Mercury really is the speediest planet in the Solar System. It hurtles around the Sun at 48 km/s, completing one full orbit in just eighty-eight Earth days. Stargazers who manage to spot Mercury one night will find it has already slipped away and shifted position only a few nights later.

All the planets orbit the Sun on roughly the same flat plane, known as the ecliptic. We can think of the planets as cars, each travelling in their own lane around a roundabout. From our vantage point aboard Earth-on-wheels, the other cars move against a fixed backdrop of buildings. The quicker the car, the faster we see it traverse that landscape. In the Solar System, the 'buildings' framing the ecliptic plane are a belt of twelve familiar constellations collectively known as the Zodiac. Sagittarius, Taurus, Leo and so on form the scenery against which we measure planetary motion.

On our roundabout, we can imagine that lanes further out have a lower speed limit. Further from the Sun, where its gravitational pull is much weaker, planets don't require as much speed to remain in orbit and avoid falling towards it. Mercury, in the orbital 'lane' closest to the Sun, takes a week or less to cross a single constellation, often overtaking planets such as Jupiter, which will saunter across the same background stars over the course of several months.

The blistering pace that Mercury enjoys as a result of its proximity to the Sun comes at a price: a blistering solar wind. Mercury's meagre gravity and the Sun's relentless wind conspire against it, carrying off its lighter elements until nothing resembling an atmosphere remains. What is left is a rocky world just over a third of the diameter of Earth. In terms of living space, it would get crowded quickly, but that is not going to be a problem seeing as it is far too hot for life and, even if it wasn't, there is no air to breathe. In addition, the lack of protective atmosphere has left the surface pockmarked by craters, evidence of a relentless beating from asteroids large and small over billions of years.

A walk across its barren landscape would quickly grow

monotonous, a seemingly endless expanse of solidified lava. There are cliffs and expansive basins, but no trees, no grass, no life. A day on Mercury is so long, fifty-nine Earth-days, that the Sun appears to move across the sky at a snail's pace. If afraid of the dark, you could choose to stroll along the surface in perpetual sunlight, easily outpacing the sunset and avoiding the encroaching shadows of twilight.

Except, what's this? As we traverse the scorched surface in sunlight six times brighter than we experience on Earth, we stumble across a patch of shadow. It's as strange as finding London shrouded in darkness while all the rest of England is enjoying the morning sun. The ground has been so bare, so unremarkable, that it has been hours since we have even bothered to look down at our feet. But now we do, as we suddenly find ourselves losing our footing: slipping, sliding, skating across a frozen plateau. As we regain our footing, we struggle to reconcile the presence of this impossible ice on a planet with daytime highs over 400 °C. We turn on our torch and glance around the darkened expanse, and we remember the other traits of the Ancient Roman namesake. Mercury wasn't just a messenger; he was also the patron of trickery and mischief. And what better planetary sleight of hand than ice on the doorstep of the Sun? Mercury has delivered one of astronomy's greatest surprises, and we will need radar to get to the bottom of it.

Before radio astronomy arrived on the scene in the mid-twentieth century, scientists did not find the idea of ice on Mercury particularly outlandish, but they believed it was possible for the wrong reasons. Back in the 1880s, optical observations had (erroneously) led astronomers to believe that one side of Mercury always faced the Sun. It seemed plausible at the time because we see the same thing with our own Moon. We only ever see one side of our Moon because it is 'synchronously tidally locked' to the Earth. This means it rotates once on its axis in exactly the same time it takes to complete one orbit around Earth, twenty-seven days. If the Moon didn't rotate at all, we would gradually glimpse more of its far side as it moved through space. But because it rotates at just the right rate, the far side always stays hidden from view – slipping away as the Moon turns.

Tidal locking is not a cosmic coincidence, but a consequence of a long gravitational dance between two celestial bodies, in this case the Earth and the Moon. Originally, the Moon was spinning far faster, but the tug of Earth's gravitational force raised a slight bulge on the surface facing our planet. As the Moon rotated, Earth still exerted a gravitational force on the bulging surface, creating a dragging force and slowing the rotation ever so slightly. As a new bulge formed on the part of the surface now facing Earth, the process repeated until, over billions of years, the Moon's rotation was slowed until it eventually came to match its orbital period.

It made sense to assume Mercury would behave in a similar fashion. It is small compared to the Sun, and lies in a close orbit, similar to the Earth–Moon system. After fuzzy optical observations of Mercury in the late nineteenth century appeared to show surface features staying still night after night, just like the Moon, astronomers assumed that it was also synchronously tidally locked. If Mercury's day were equal to its year of eighty-eight Earth-days, then this would leave one hemisphere of Mercury eternally facing the Sun. Astronomers imagined a world of two halves: one a barren, scorched wasteland, and the other a cold, dark expanse. With no atmosphere, there could be no wind to transfer heat and so the two sides would remain in stark temperature contrast. The dark side of Mercury could conceivably host ice, and for the longest time Mercury was thought to be a partly frozen planet.*

* Two stories by Isaac Asimov are built around this widespread belief: 'Runaround' in *I, Robot* (1950) and *Lucky Starr and the Big Sun of Mercury* (1956).

It would take about eighty years, and the advent of the radio telescope, before astronomers realised that those early optical observations were wrong. Mercury wasn't moving as expected at all. But how did the idea of a world divided in two, one side scorched, the other frozen, persist for so long? Why did it take an antenna to challenge this enigmatic, but ultimately incorrect, picture?

There is no doubt regarding tidal locking when it comes to the Moon; anyone can see the proof. The same features – the craters, the 'man on the Moon' – are always staring down at us. In contrast, Mercury is so close to the Sun that it is only visible just above the horizon for a very short period before sunrise or after sunset, depending on the time of year. In the 1880s, when the erroneous observations were made, telescopes were simply not powerful enough to discern surface features clearly enough to see them moving. By the 1940s and 1950s, optical telescopes were powerful enough to see individual craters on Mercury in theory, but the seeing conditions (blur from Earth's atmosphere near the horizon) were too challenging. In addition, astronomers then, as now, were reluctant to risk their telescopes' sensitive instrumentation. Even the great Hubble Space Telescope doesn't dare cast its eye on Mercury. The bigger and more impressive the telescope, the greater the risk.

Radio astronomers, on the other hand, have no problem pointing their antennas in the direction of Sun. The Sun does produce radio light, as discussed in the previous chapter, but only relatively weakly. Even in the strongest solar storms, radio telescopes are largely immune to permanent damage: a cheap and replaceable electrical component fried once every decade or so perhaps. These storms are rare, so the risk to a radio telescope when looking for Mercury is negligible, in comparison to the permanent damage that would be wrought on the Hubble or James Webb Space Telescopes if they risked a glance.

In 1962, the nascent explorations of radio astronomy began to cast doubt on the theory of Mercury's tidal locking.[1] First, the radio astronomers just listened, observing the natural radio light Mercury was emitting from its heated surface: its thermal emission.

One would expect the side facing the Sun to emit high levels and the opposite side to emit extremely low levels. Yet the radio observations showed something puzzling. The supposed 'dark side' of Mercury was far too warm.

Astronomers decided to probe this oddity a little more directly than just by listening for thermal emission. They turned to radar. Mercury was too far for radio astronomers to produce maps, but they could instantly tell how fast Mercury was rotating by using the Doppler method.

The best way to understand this most important of radio astronomy methods is to imagine a fire engine passing by, and note how the pitch of the siren seems to get higher as it comes towards you and then lower again as it drives away. As the fire engine approaches, it has moved a little closer to us each time the siren reaches its peak. Successive sound waves therefore arrive at our ears in shorter intervals, leading to a kind of 'pile-up'. The distance between the waves as they enter our ears, the wavelength, is decreased, which we hear as a higher frequency, or pitch. Conversely, as the fire engine moves away from us, the sound waves are emitted with progressively longer intervals, increasing the wavelength, which we hear as a lower pitch. We can even get an intuition for how fast a fire engine is moving by how drastically the pitch of the siren is changing.

We can apply the same idea to light waves. Radio waves reflected from the side of the planet rotating towards us appear slightly shorter in wavelength ('bluer'), and the waves bouncing from the edge of the planet rotating away from us appear at slightly longer wavelengths ('redder'). The greater the change in wavelength, the faster that planetary surface is moving. The

Doppler method is still employed by every astronomical radar facility out there, because it facilitates the near-instantaneous measurement of planetary surface velocities.

The Doppler method can be applied to any kind of light, but the magnitude of the shift increases with longer wavelength, making observations much easier if you use radio waves. Optical Doppler measurements demand extraordinarily precise instruments, even by today's standards, because their wavelength is thousands of times smaller than the width of a human hair. The Doppler shift induced by Mercury's rotation would be considerably smaller than that, a change so slight it amounts to just a few trillionths of a metre in wavelength. By contrast, radar Doppler methods work with wavelengths on the order of centimetres or larger, resulting in shifts of about a nanometre. Small, but well within reach even for the first generation of radio telescopes, of which the great Arecibo was one.

Arecibo Observatory is close to the northern coast of Puerto Rico and was conceived in the early 1960s not as an astronomical facility, but foremost as an atmospheric research centre. At the time, the US government was investing heavily in research to detect the movement of nuclear warheads. Physicist, astronomer and electrical engineer William Gordon calculated that one could build a 305-m-diameter radio dish to collect the radio light scattering from the trails of missiles travelling through the upper atmosphere; $10 million landed in Gordon's lap. It was only during construction that Gordon realised he had grossly overestimated the diameter of the dish required – a 30-m dish would have done it. But it was a mistake that greatly benefited astronomy for over half a century; no other radio telescope has been close to competing with its power. Had Gordon rectified the scale, I'm not sure when astronomers would have been able to successfully secure funding for the full 305 m: $10 million for a nuclear threat detection system is one thing, but I doubt there would have been such ready funding available to probe the shadows of Mercury.

When complete, it looked like a smoothly polished crater in the landscape, a clean, slick, metal construction covering twenty acres and incongruously surrounded by a tangle of jungle and hills. (This is the kind of jungle one imagines Tarzan swinging through: vines really do hang down from trees, geckos scramble across every surface and minuscule tree frogs make a deafening noise.) Manufactured 'vines' hung from the support towers too. A series of thick, steel cables stretched down toward a platform held 150 m above the dish, holding the 'listening' part of the apparatus, the antenna. This antenna could both listen and, crucially, send radio waves. Less than two years after its construction, the Arecibo sent a powerful beam of radio waves to Mercury.

In 1965, these Doppler radar observations detected a slight shift in the reflected radio waves from Mercury, revealing that it was indeed rotating, but not in sync with its eighty-eight-day orbit, as had been expected from those early optical observations.[2] Instead of completing a rotation every eighty-eight Earth-days, Mercury spins on its axis every fifty-nine days – faster than expected. Rather than being tidally locked, turning once per orbit like the Moon does around Earth, it turns three times for every two laps around the Sun. This discovery was not merely a technical correction; it was the first clue that tidal forces could affect planetary rotation in surprising ways. It forced astronomers to consider what it was that was causing Mercury to behave in this unexpected way.

The key lies in Mercury's elongated orbit: unlike Earth's nearly circular path, Mercury follows a distinctly stretched ellipse. In such an eccentric orbit, the Sun's tidal grip can stabilise Mercury in the unusual 3:2 rhythm rather than the more familiar 1:1 lock. But why should Mercury's orbit be so elongated compared with the nearly circular orbits of all the other planets of the Solar System? Many planetary scientists now see that eccentricity as a scar of Mercury's violent early history. Models suggest that Mercury actually formed further out from the Sun, before giant impacts and gravitational jostling with other planets caused an inwards migration, stretching its orbital path and leaving it trapped in its current spin-orbit dance.

This highly elliptical orbit causes the planet to move much faster when it is closer to the Sun, such that its orbital movement can temporarily outpace its own rotational speed. One could sit down on Mercury to enjoy the sunrise, only to see it stop before it was entirely overhead and, amazingly, slowly reverse back across the sky. From some points on the surface, the Sun actually sets in reverse, until Mercury moves far enough from the Sun that its orbital velocity falls below the rotation speed. The Sun rises overhead once again, a double sunrise, finally completing a 'real' sunset eighty-eight Earth-days after that first sunrise.

What with the Sun darting back and forth across the sky, it is not surprising that the surface of Mercury has a complex temperature distribution. Much like the meandering path of the Sun across the Mercurial sky, astronomers have been on quite the journey of understanding. Prior to the 1880s, the idea of ice on a world so close to the Sun was absurd. Then, following erroneous observations suggesting Mercury was synchronously tidally locked, astronomers settled in to seeing Mercury as a half-icy world. Then, in the 1960s, radar data instantly shattered the idea of a frozen hemisphere without ceremony. A new dogma was established: Mercury, for all its weird and wonderful motion, was most probably an exclusively scorched planet. How to be sure? Perhaps a visit might be in order.

In the opening paragraphs of this chapter, we took a stroll across Mercury, a stroll which is, of course, purely fantasy. No living creature could withstand the planet's extremes: daytime temperatures soaring above 400 °C, frigid nights plunging to −180 °C, a world without air or liquid water, bathed in relentless solar radiation. No human has ever set foot on Mercury, and no robot either. There have been no landers, rovers or even the tiniest probes. This is because landing on or even orbiting Mercury presents an extraordinary challenge, one rooted in the immense effort required to shift from Earth's path around the Sun to Mercury's, and then keep pace with the planet.

In 1974, *Mariner 10* became the first mission to pass nearby to Mercury. It was designed to orbit the Sun, an easier task, and skim past Mercury three times as it did so. On this 'fly by' mission, *Mariner 10* snapped pictures with its optical camera as fast as it could during each of its three observation opportunities – visits lasting only hours. On this planetary safari, the spacecraft acted not as an all-terrain jeep with the freedom to roam, but as a tram with a fixed route, strict timetable and no stops. It was a fleeting, controlled expedition, only catching glimpses of what Mercury had to offer.

Mariner 10 was able to capture images of about 45 per cent of Mercury's surface and, as beautiful as the photographs were, they aligned with our expectations: a cratered, desolate planet, and nothing more. It wasn't until several decades later, in 1991, that astronomers discovered there were corners of Mercury hidden from optical sight.

On Earth, every part of the surface eventually experiences sunlight due to the axial tilt of 23.5 degrees. This tilt leads to our familiar seasons and variations in daylight hours. When a hemisphere is tilted towards the Sun it will experience summer with longer days and shorter nights. At the equator, the variation in daylight hours is minimal, but even at the North Pole, where the polar night brings months of darkness, the Sun inevitably rises again.

Radar data doesn't just give us the rotation speed of a planet, but also tells us on which axis it is rotating. Draw a line between the half of the radar map that is redder (moving away from us) and the half that is bluer (moving towards us) and you have the axial tilt of the planet. Unlike Earth's generous tilt, radar data clearly shows that Mercury's is almost zero. This means that while most of its surface gets baked, parts of its polar regions remain in permanent shadow, especially within the deepest craters.

When we map a planet's surface with optical photography, we rely on reflected light from the Sun. So if a part of the surface is hidden from the Sun, perhaps by the walls of a crater, it's hidden from us too. Artificially illuminating these portions of Mercury from a space probe would require impossible amounts of energy. But radio waves occupy the lower-energy part of the electromagnetic spectrum, so a radio-light source requires far less power, offering a solution. With an antenna and a modest power source, you've got a radio camera equipped with a 'flash', capable of illuminating every crevice and crater, heedless of the shadows.

In 1991, the 70-m antenna at NASA's Goldstone Deep Space Communications Complex in California sent out radio waves to Mercury for two days, while over two dozen 25-m antennas constituting the Very Large Array (VLA) in New Mexico listened intently for the echoes.[3] We had set off on a planetary safari like on *Mariner 10*, but this time with a more sophisticated toolkit. Instead of simply pointing the camera and hoping to spot something interesting as our tourist tram trundled past, we used radar waves to feel our way around Mercury's hidden landscapes, mapping the areas of Mercury that *Mariner 10* had not reached. Like a team of tour guides spread across the plains, these radio waves picked up the subtle signs and changes in the landscape that might indicate a place of interest.

Much of radar work is usually done with just one antenna, switching between transmitting and receiving modes. If you have ever used a walkie-talkie, you will be familiar with one limitation of this 'monostatic radar': you cannot hear the other person while you are pressing the button to transmit. However, using two (or more) antennas in tandem to transmit and receive, 'bistatic radar', allows one to transmit and listen continuously. In addition, just as two eyes give us depth perception, and two ears allow us to pinpoint the source of a sound, two radio telescopes help us work out the exact origin of the radar echoes. With antennas placed in different locations, we catch the radio waves that glance off surface features at different angles.

There was a great deal of surprise that the radar maps of Mercury had some very bright patches, especially over the planet's north pole, indicating the presence of a material that reflects radio waves particularly well. The surface of Mercury, like the Moon, is covered in a fine layer of regolith, dust and crushed rock formed by billions of years of meteorite impacts. This regolith, and the rocks beneath, reflect plenty of radio waves but they absorb some of them too, lowering the intensity of the reflected light. Ice, unlike regolith, reflects radio waves with such brilliance that the returning signal dazzles like sunlight glinting off fresh snow. One cannot underscore the astonishment this result caused: it had been almost three decades since anyone had seriously considered the idea of ice on Mercury. It wasn't just the brightness that gave the game away, though; it was the polarisation of the light too.

If you flick a rope up and down, the wave travels forwards, but the motion of the rope is vertical – this is like a *linearly polarised* light wave, where the electromagnetic field vibrates along only one direction. If you instead lift the rope and swivel your hand in a circle, the rope rotates and traces a spiral path – this is like a *circularly polarised* light wave. Sunlight, however, is completely *unpolarised* – its electromagnetic field vibrates randomly in all directions. If we transmit radio light of a known polarisation, the polarisation it returns with tells us about the surface it has reflected off. For example, materials such as regolith change the polarisation of radio waves, while ice does not.

The polarisation data wasn't merely a redundant confirmation of the locations where the radar brightness had indicated ice. There were also areas where the brightness was lower, yet the polarisation remained largely unchanged. This indicated the presence of ice, but ice that was partially buried under a dusting of regolith. The depth of the underground ice can be estimated by the degree to which the light is changed in terms of its brightness and polarisation. There are ice sheets up to 10 m thick in places, often covered in about 10 to 20 cm of dirt. Radio waves sent from the Arecibo Observatory quickly evaluated every part of Mercury and estimated that it contains between 100 billion and 1 trillion t of ice.[4] This is only equivalent to a single mid-range glacier on Earth, but this is Mercury we are talking about – the closest planet to the Sun! The existence of ice in any amount is counter-intuitive and awe-inspiring in equal parts.

We are inclined to think of Mercury as a hot, barren planet, but that perception is misleading. Once we resist the urge to simplify everything into neat boxes, hot or cold, black or white, we can choose to be led only by scientific curiosity and discover realities we would never have expected. The radio data from Mercury is unequivocal: hidden away in craters at the poles, protected from the Sun's relentless rays, ancient ice sheets have quietly gathered dust. Once we have absorbed the fact that ice has been found on Mercury, we might find ourselves asking: how does it get there?

On Earth, even in the small space of your kitchen, we easily find water in its three forms: ice in the freezer, steam billowing from the kettle, and liquid water flowing from the taps. Water in any form is made of just these basic ingredients: two atoms of hydrogen and one of oxygen. Hydrogen is the simplest of atoms, formed in such abundance following the Big Bang that it still constitutes 73 per cent of the mass of the visible Universe today. Oxygen is heavier and, like many heavier elements, is formed within stars before finding itself expelled into space during the supernova marking the end of that star's life.

The Sun only formed 4.5 billion years ago, from a dusty gas cloud already enriched by supernovae, so both ingredients were readily available. However, they were not equally available across the nascent Solar System. While they may have begun as equally distributed, the intense solar wind of our young Sun blew the lighter elements outwards. Planets forming close to the Sun do so in an environment where light elements are rarer. For a small world like Mercury, with meagre gravitational pull, these sparse gases easily escape into space, stripping it of the chance to form an atmosphere. Without an atmospheric shield, Mercury could not sustain surface water, even in the sheltered shadows of its poles. To explain ice on Mercury, the water has to originate further out in the Solar System.

In mythology, Mercury is the swift-footed messenger of the gods, racing between worlds to deliver news from the heavens to Earth. But when it comes to the cold case of ancient ice on a hot planet, Mercury played the role of recipient. In the distant past, it appears he was the beneficiary of a package from outer space, delivered by asteroids and comets.

In this book's dedication I allude to one of the most formative moments of my childhood: seeing the comet Hale–Bopp streak across the night sky. For months its bright nucleus could be seen shining in the night sky, leaving a tail of dust and gas. Comets, often described as 'dirty snowballs', form in the distant, icy regions of the Solar System. These frozen bodies are made of volatile compounds like water, ammonia and carbon dioxide. As a comet approaches the Sun, its icy surface sublimates, transforming from solid to gas, creating a glowing tail that was so beautiful it made me cry.

Asteroids, by contrast, are rocky remnants left over from planet formation in the inner Solar System. They drift in the asteroid belt between Mars and Jupiter, where ices cannot form so readily. Asteroids tend to contain less water than comets, but some, particularly a type known as carbonaceous chondrites, are rich in hydrated minerals. These asteroids carry water locked within their structure in the form of clays and salts.

These meteorites fall all over our planet and have done for billions of years. It's thanks to these collisions early on in Earth's formation that vast oceans cover the surface of our planet, the water released from within asteroids and melted from within comets filling the basins of our crust. Earth was too close to the Sun for water to have formed naturally in large quantities, though our atmosphere may have allowed a small amount to form. We tend to think of asteroid collisions as causing the end of life, thanks to the cautionary tale of the dinosaur extinction. But, without asteroids bringing water to Earth in vast quantities, our planet would have been a barren wasteland. More than that, the rarest of asteroids contain amino acids, the building blocks of life, and it is thought that the formation of life on earth depended on these cosmic shipments.

Mercury, without an atmosphere to protect it, is subject to an even more intensive bombardment, but water cannot survive for long on most of the surface, sublimating at the first touch of searing sunlight. Any ice delivered directly into regions in permanent shadow could remain frozen indefinitely, though. While on Earth these ices melted to form the oceans, and incubate life, on Mercury whole slabs lie in the deep darkness, just where they landed.

Mercury has been relentlessly pelted by space debris, and each impact stirs up dust, spreading the fine regolith across the surface. It settles everywhere, filling craters, coating rocks, and even blanketing the ice deposits hidden in shadowed regions near the poles. The result is a layer of grime that dulls the brilliance of the ice, much like that on snow piled by the roadside. On Earth, the blacker the snow, the longer it's been there . . . or the heavier the traffic. On Mercury, it's a similar story. By counting craters, scientists can estimate the rate of bombardment and, from that, work out how long it might take for a certain depth of regolith to accumulate over ice. Radar observations of dusty ice serve as a kind of timestamp, letting radio astronomers estimate how long ago the ice was delivered. The radar observations from Mercury reveal a level of regolith coverage suggesting that the ice has lain on Mercury for hundreds of millions of years.

To summarise, twentieth-century radio astronomy delivered two definitive revelations regarding the fastest planet in the sky. First, the earliest purpose-built radio telescopes pinned down how fast Mercury spins on its axis and revealed it has no 'dark side'. Then, three decades later, the Arecibo 305-m dish showed that around the shadowy poles lay ice that had not seen the light of day since it was transported there by asteroids and comets.

For humans, seeing (in optical light) is believing, however, and we couldn't resist making sure. In 2011, twenty years after the radar observations, the *MESSENGER* probe confirmed the unmistakable signature of water ice.[5] A remarkably recent date, considering the first full-surface radar images were returned decades earlier, and the last visit to Mercury was the *Mariner 10* mission, which ended in 1975.

The challenge of landing on Mercury is not one that has been solved in the time since *Mariner 10*. Think of the Solar System as a racetrack where the inner lanes demand the fastest pace. Mercury, on the innermost lane, is sprinting. If you just cut across without adjusting your stride, you'll overshoot. If a spacecraft simply falls inward from Earth's orbit, it ends up on a stretched path that shoots it past Mercury rather than keeping pace. To be captured, it must shed enormous energy to reshape its orbit and fall into step with Mercury's motion. Using thrusters alone to induce this change would require an unfeasible amount of fuel, so missions instead take a winding route, using repeated fly-bys of Earth, Venus and even Mercury itself to gradually bleed away energy and bend the spacecraft's path so that, little by little, it begins to move in step with Mercury.

The *MESSENGER* probe used these 'gravity assists' at Earth, Venus and around Mercury until it was moving in the right direction and speed to be captured permanently by Mercury's gravity. But this is not a quick process – it is faster to travel to Jupiter, a planet ten times further. *MESSENGER* took 6.5 years to reach Mercury, during which the scientists could only watch with terror as each gravity assist posed a new, very real danger to the spacecraft. In comparison, radio light takes just over six minutes to probe Mercury's craters and return with the information.

And as for landing on the surface, where radio photons have already dived into shadows and under regolith, there is just one sign of physical exploration on Mercury. One crater among the many that litter the surface was formed by the impact of the decommissioned *MESSENGER* probe in 2015. As it reached its fuel reserves, NASA ordered it to use that last fuel to set a new trajectory to the surface of the planet it had watched over for four years. It collided at 14,000 km/h, and its crumpled remains still lie on the surface in an eternal resting place. At least it won't be alone for long. *BepiColombo*, the third and latest mission to Mercury, will settle into orbit in late 2026, following its launch in 2018.

With a new probe soon to be in orbit around Mercury, and three decades on from the radar discovery of ice, one might wonder if radio astronomy has anything left to offer when it comes to the exploration of Mercury. The answer, in short, is yes.

Even when safely in orbit, a probe must still use small bursts of fuel to propel itself against the constant, clawing gravity of a hungry Sun. This severely limits mission lifetimes, making them much shorter in comparison to the travel time. So, while the presence of a space probe around Mercury is to be celebrated by all astronomers, it is in the back of our minds that it will be a handful of years at most before it plummets towards the hot, icy, mercurial surface.

It is always shocking to me, the ratio of travel to mission time, and how quickly the names of these missions are forgotten. There are so many of them across the Solar System, each lasting only months, maybe years. As a radio astronomer, I am used to encountering the same radio telescopes woven throughout the entire history of my research field, right to the present day: Arecibo, Goldstone, Green Bank, to name just a few.

Radio telescopes are engineered for durability and adaptability: every nut and bolt can be replaced; the surface panels updated to more reflective materials; the antennas connected to new super-computers.

As long as there is a modest amount of funding, radio telescopes can adapt, pivot and transform into evolved versions of their original selves, able to produce groundbreaking new science over timescales spanning generations of scientific careers. It is a sustainable form of exploration, all the more valuable when contrasted with the difficulty of sending a probe to visit Mercury.

There are no planned missions to Mercury following *BepiColombo*, but should a new idea be mooted, it will face the usual hurdles of budgets and politics and require a lot of patience. Meanwhile, Earth-based instruments like the Green Bank Telescope are continually sharpening their planetary radar capabilities. And, as always, we're not limited by rocket launches, mission windows or the political pressures that come with funding entirely new scientific endeavours. We can choose to explore the innermost planet whenever the fancy takes us.

For now, though, Mercury, the swift-footed messenger of the gods, has delivered a story etched in dust and shadow. Now the torch passes to Venus, where the truths we seek lie deeper still, hidden beneath a thick and unyielding veil of cloud.

4

VENUS

> Next time you meet a climate-change denier, tell them to take a trip to Venus; I will pay the fare.
>
> Stephen Hawking, *Stephen Hawking's Favorite Places*, Episode Two

If our Solar System had a guidebook, Venus would be the ultimate 'off-the-beaten-track' destination: intriguing yet utterly inaccessible to the casual tourist. Unlike the Moon, where we can easily see craters and plains with our own eyes, Venus hides behind a thick, opaque layer of clouds. Yet astronomers are the most intrepid of cosmic spies. No matter how daunting or dangerous the mission, they will find a way to infiltrate. Thanks to the ingenuity of radio astronomy, we have peered beneath the cloud curtain encompassing Venus, uncovering a world so hostile that it shattered all our romantic associations with our closest neighbour.

As one of the inner planets, Venus is always relatively close to the Sun in the sky. While it orbits the Sun, it appears to the east of the Sun half the time, and to the west the other half. This means Venus either rises just before the Sun or sets just after the Sun. When it passes directly behind or in front of the Sun, it disappears altogether. Over the course of about nineteen months, Venus gently trades one persona for another, from 'morning star' to 'evening star'.

Its bright presence and calm, predictable movements led ancient societies around the world to equate the planet with love, peace and beauty. Venus was the Roman goddess of those attributes, but the Egyptians, Greeks and Babylonians named the planet after their equivalent goddesses: Isis, Aphrodite and Ishtar, respectively. So synonymous with womanhood is Venus that the planet's astrological symbol (♀) is also used as a symbol for the female gender. Indeed, it feels odd to refer to Venus as 'it', rather than 'she'.

Venus is considered our sister planet because it is similar in size and, as second planet from the Sun, not too far away from our own position in the Solar System. Up until the late 1950s, the possibility of life on Venus was not considered out of the question. Science-fiction writers and scientists alike pondered what landscapes might lie beneath the clouds. In 'All Summer in a Day' (1954), Ray Bradbury wrote of a group of schoolchildren on Venus who had no memory of anything but unbroken clouds and perpetual rain, exploring how humankind would overcome the psychological challenges.* Meanwhile, in *Lucky Starr and the Oceans of Venus* (1954), Isaac Asimov imagined a global ocean world, with an underwater civilisation akin to the Lost City of Atlantis; and C. S. Lewis wrote of a planet of floating islands populated by a diversity of lifeforms in *Perelandra* (1943). The mythos of Venus has even touched the non-fiction shelves, in

* Had he lived in England, he would not have needed to imagine at all.

John Gray's popular-psychology book *Men Are from Mars, Women Are from Venus* (1992). This book argues that women are inherently peaceful and gentle like the planet Venus, a notion unsupported by biology, astronomy or everyday life.

Radio astronomy, however, has shattered millennia of misconceptions, showing that it is entirely inhospitable to life. If sci-fi films such as *The Martian* are anything to go by, all you need to survive as a man on Mars is a radiation suit and some potatoes. As a woman on Venus, you'd need far more than that. It is scorchingly hot, hotter even than Mercury, and its atmospheric pressure is over ninety times that of Earth. To survive any time at all there, one would have to be as hard as nails. Except harder, because nails would crumble to pieces in the 475-°C heat. Peaceful and gentle, Venus is not.

Even sending uncrewed probes has resulted in numerous setbacks and expensive failures. There is no other planet in our Solar System so hostile, even to light. The clouds stop optical photons in their tracks, and while X-rays can reveal our own skeletons easily enough, the skeletons in Venus' closet remain resolutely hidden. Infrared, too, usually used to track lifeforms hiding in the darkness, fails to illuminate the surface. Yet our drive to comprehend how two sister planets could evolve so differently remains as strong as ever. Luckily, we have other ways of slipping through planetary borders undetected and unharmed. Radio light can infiltrate this planet's atmospheric defences, returning with maps of the terrain and intel on the turbulent volcanic forces at work. Through these covert channels, we continue to unlock Venus' secrets, using its own radio emissions and our radar capabilities to get to know our most isolated and impenetrable neighbour.

Every form of light has its superpower, whether it be cooking food (microwaves), exposing bones (X-rays) or long-distance communication (radio). Infrared light is the thermometer of the group, and so the go-to tool when we wish to measure the temperature of an object, whether it be a hiding fugitive or the surface of a planet. When matter is heated, it produces thermal emission across the spectrum, with the temperature determining the most favoured wavelength.

You can see this with a simple experiment: stick a poker in a fire, and it glows red hot because, in matter heated to over 500 °C, the thermal emission peaks at visible red wavelengths. Humans, at a much cooler 37 °C, don't glow like a red-hot poker because our thermal emission peaks in the invisible infrared range. That's why we need infrared cameras to detect body heat in the night, but don't need one to see the glowing poker.

If we measure the amount of light at a variety of wavelengths, we can match the pattern of emission to a specific temperature. It is kind of like trying to match a colour swatch with a paint tin. The thermal emission curve acts like a unique colour blend, and by comparing it to our chart of known paint shades, we can identify the exact hue or, in this case, the temperature.

Today, infrared astronomy is generally carried out from space, but when astronomers were first trying to study the planets, they made crude infrared observations from high-altitude observatories. Infrared readings of Venus in the 1940s and 1950s suggested average temperatures only a few degrees above zero, hinting at a potentially habitable environment. Imaginations went wild. However, in 1956, radio astronomers took the scorching reading of around 475 °C.[1] This was not just a case of a planet heating up in the day and cooling down at night – both readings stayed the same whatever the time of day. The scientific community was shocked and confused. How could there be two such vastly different temperature readings?

The long-held, baseless preconceptions of a serene sister-planet teeming with life contributed to a collective unwillingness to accept this new reality. The astronomers had to defend their measurements to their colleagues, to the press, and even to an angry member of the public who rang to give them a piece of his mind. Both thermometers could not be correct, and if the radio reading was the one to trust, that high temperature meant our sister planet was empty of life. This was too hard to stomach, and denial set in. Everyone knew temperature was infrared's superpower, after all; why should they trust the radio astronomers?

When NASA initiated the Mariner programme in the early 1960s, Venus was their first priority, partly to settle the temperature dispute, but mostly because it was still thought to be the planet most likely to host life. *Mariner 1* had failed on launch, but NASA's *Mariner 2* mission flew by Venus in 1962, measuring its thermal emission at multiple wavelengths. To everyone's consternation, the temperatures were exactly as measured on Earth: the infrared readings indicated Venus was enjoying a perpetual autumn day, while the longer wavelength observations were more reminiscent of Dante's *Inferno*.

This discrepancy, astronomers had come to understand, was due to Venus' dense atmosphere. The infrared light was coming from Venus' cooler cloud tops, while the radio waves were emitted by the scorching surface. Both the surface and atmosphere emit thermal radiation over the electromagnetic spectrum, peaking in the infrared. Infrared light from the top of the atmosphere, where it is cool, escapes easily into space and a probe like *Mariner 2* measures a clear peak corresponding to those milder temperatures. The infrared surface emission, however, is scattered and reflected by the clouds, never making it to any telescope. In contrast, the much fainter thermal emission in the radio wavelengths passes through the atmosphere with ease, providing a clear temperature reading of the searing landscape below. There are some environments only radio can traverse. Its superpower is not just long-distance communication, it is communicating with the otherwise unknowable.

The *Mariner 2* observations confirmed that, although Venus' atmosphere appeared to be a temperature conducive to life, the planet's surface was far hotter, exceeding even that of Mercury, a planet nearly twice as close to the Sun. But Mercury, if you remember, does not have an atmosphere. So, as much as it heats up quickly in the daylight, it also cools down quickly at night. On Venus, the atmosphere is dense and contains a lot of carbon dioxide, a potent greenhouse gas. These gases act as a mirror to thermal radiation, trapping heat with brutal efficiency. The planet heats up, causing it to produce more thermal radiation, which in turn cannot escape. This vicious cycle of increasing temperatures is only made worse by the intense volcanic activity throwing more greenhouse gases into the atmosphere. Once hailed as a potential second Earth, Venus was proven to be an uninhabitable pressure-cooker. And radio astronomers, listening to the whispered secrets of a lifeless planet from over 100 million km away, had been the first to know.

Sending a probe like *Mariner 2* to another planet requires pinpoint precision in calculating distances and orbital dynamics, since even a tiny error can lead to one missing the mark by millions of kilometres. By timing how long it takes radar waves to return from Venus, astronomers have been able to measure the movement of Venus with extraordinary accuracy, ensuring that space missions are aimed not where Venus is now, but where it will be when the probe arrives. This is a crucial exercise, as space probes do not have enough fuel to cope with a diversion. If you get it wrong, mission failure.

Venus was the first planetary body we attempted to reach by radar, primarily because it is the closest planet to Earth by a significant margin. Still, at over a hundred times further away than the Moon, detecting an echo from Venus was an immense challenge. Because the strength of a light signal gets rapidly weaker with distance, an echo from Venus would be about *10 million* times fainter than an echo from the Moon. Technology had to catch up with these ambitions, leading to a fifteen-year gap between the radar detection of the Moon in 1946 and that of Venus. Finally, in 1961, first contact was made, resulting in a distance measurement to Venus so much more accurate than any other estimate that it resulted in a redefinition of the Astronomical Unit (AU), the standard measure of distance in space.[2] While the AU has been redefined since, the 1961 estimate was correct to about 0.0001 per cent, equivalent to measuring the distance from London to Sydney and being off by the width of a football pitch.

During the backdrop of cold war, the Soviets chose to conduct an experiment to test the power of their telegraphic communication systems, similar to the moonbounce experiments. Using radio waves they transmitted the word *MIR*, Russian for 'peace' and 'universe', as one of their first telegraphic transmissions to Venus, as well as 'USSR' and 'LENIN'.[3] Sure enough, five minutes after transmission, the echoes of these words were gathered by the listening radio antenna. Presumably, if the US equipment was listening at the time they would have detected these signals easily enough, too. I sometimes wonder if there were messages bounced off Venus, or the Moon, that did not make it into the history books. One can imagine two rooms of radio scientists, separated by geography and political ideals but, half-bored and half-curious, exchanging insults via Venus.

Before we first sent radio waves to Venus, we didn't even know where the planet's rocky surface ended and its atmosphere began. By eye, our only estimates of Venus' size are estimates of the size of its atmosphere, because we cannot see underneath. If we flood Venus with radio waves, only the surface of the planet will reflect them; the radio waves which skim past the planet's surface will go straight through the atmosphere and career off into space. The resulting circle of reflected radio waves reveals Venus' solid surface diameter to be just over 6,000 km, about 500 km narrower than optical measurements suggest. This suggested that the atmosphere of Venus extends for up to 250 km, twice what we experience on Earth, and uncrewed probes later confirmed the radar estimate.

That same burst of radio waves gives us more information for free, if we know what to look for.

Consider the task of gauging the length of a day on Venus. For Jupiter, the challenge is less daunting: its colossal red spot serves as a beacon, a moving marker with which we can time its rotation; Mars has its ice caps; even the Sun has its spots. Venus, however, is a featureless sphere to the eye. Whether spinning or standing still, it looks the same. In radio, though, the Doppler effect leads to a measurable shift in the wavelength of the reflected light, proportional to how fast the surface is moving. The magnitude of this shift gives us a direct estimate of the speed of the rotating surface, and from that we can easily work out the rotation period, i.e. the length of a day. With just one burst of radio waves, we can discern the length of a Venusian day, 243 Earth-days, bypassing the need for any fixed landmark.

This is by far the slowest rotation of any Solar System planet, so slow that its year (time taken to complete an orbit of the Sun) is eighteen Earth-days shorter than its day (time taken to complete rotation on its axis). If celebrating your eighteenth birthday on Venus, you would technically only be sixteen Venus-days old (and only eleven in Earth-years). What's more, radio Doppler data reveals that a Venusian inhabitant would observe the Sun rising in the west and setting in the east. Venus is spinning the wrong way.

Venus is rotating on its axis in a retrograde motion, which is clockwise if viewing the Solar System such that you are looking down onto the North Pole of Earth, and opposite to its anti-clockwise orbital motion around the Sun. The 'normal' motion for Solar System planets is prograde motion, like that of the Earth, where the planet is rotating on its axis in the same direction as its orbit around the Sun, i.e. both counter-clockwise from our bird's-eye view of the Solar System.*

Prograde motion is the expected behaviour in any planetary system because everything in a star system forms from the same spinning gas cloud. As this cloud collapses under gravity, angular momentum is conserved, meaning the original spin of the cloud dictates the direction of rotation and the orbital motion for the planets that form within it.

Something must have happened to Venus to disrupt its natural motion, like a fairground operator giving your teacup a push to spin it in the opposite direction. One possibility is that a cataclysmic collision in the distant past struck Venus on one side. We know from how our own Moon was formed that such collisions did happen in the violent early Solar System. But there's also the possibility that Venus' own turbulent atmosphere is altering the motion of the planet within, as though the teacup riders themselves have seized the wheel and are dragging it in the opposite direction. Only radio waves sent from Earth can confirm whether this theory holds true.

The Venusian day was successfully measured with radar in 1963.[4] Yet when it comes to pinning down the length to the nearest minute and second, estimates have varied ever since. It may seem fairly irrelevant whether a Venus-day is 243 Earth-days and 2 Earth-minutes or 242 Earth-days and 50 Earth-minutes. After all, there is no one down there to care particularly if we are late for a meeting. But the 'day' is a measure of rotation speed and translates to a surface moving faster or slower than we might expect.

* Uranus is the only other exception, as it revolves on its side relative to its orbit of the Sun. The Sun rises all over the place during its eighty-four-year orbit, including from the north and from the south.

This really matters when it is a surface on which a spacecraft is trying to land.

Imagine a team of secret agents skydiving onto an aircraft carrier in enemy territory. They jump from their plane having made an assumption about how fast the ship below is moving. If the aircraft carrier is moving faster or slower than they expect, they could easily miss the boat entirely, and land in shark-infested waters. A spacecraft heading to Venus might land kilometres from its chosen site.

Scientists sought to find a pattern in how Venus' day varied, but that required long-term surveillance, something beyond the reach of space missions which tend to either fly by their target, or orbit only briefly. But with ground-based radio facilities, there are no time constraints imposed by fuel, aside from a regular supply of tea and biscuits required by the astronomers in the control room. Radio astronomers can afford to be patient and keep one eye on the data trickling in while they get on with solving other mysteries.

Between 2006 and 2020, astronomers based at the Goldstone Deep Space Communications Complex in California used their 70-m dish to ping radio waves off the surface of Venus whenever they weren't busy tracking and relaying messages to NASA space missions. The echoes revealed that the Venusian day varied by as much as twenty-one minutes, and not in any predictable pattern.[5] Earth's day, in comparison, varies by less than a millisecond. All of the previous estimates, despite their apparent disagreement, were within the range of daytimes Arecibo measured. It wasn't the people taking the measurements that had been inconsistent; it was the planet itself that couldn't settle on a number.

The atmosphere, it turns out, is the culprit. The extraordinarily dense atmosphere is so hefty that it can slow the rotation of the planet, making the days quite literally drag. This might even explain why, at the end of the Venusian day, the Sun sets in the east. If Venus started out with prograde motion, it may be that the atmosphere helped to slow and reverse that rotation to the retrograde motion we observe today. We may not need to look at violent tussles with other protoplanets to explain Venus' retrograde rotation. Perhaps Venus is a planet simply at war with itself: a rocky surface battling against an atmosphere throwing its weight around. This is a problem for anyone hoping to land a probe on Venus, as any space missions would need to accept the risk that their landing zone would be moving unpredictably beneath their feet.

As it turns out, aiming for the right spot on Venus is the least of anyone's worries – the planet seems determined to annihilate anything long before it touches the surface.

In the very earliest days of the space race, there were two main goals: the first was to land a man on the Moon, and the second was to explore Venus. The clash of political ideologies between the US and Russia could not have been starker, and if a cold war had to be fought, the frigid expanse of space was a fitting arena. Where in space almost didn't matter. This was a proxy war, an effort to prove scientific and technological superiority and scare the other side away from the nuclear red button. It was all about being first, and Russia got off to a strong start:

First artificial satellite: Russia, 1957

First dog in space: Russia, 1957

First uncrewed (crash) landing on the Moon: Russia, 1959

First man in space: Russia, 1961

First woman in space: Russia, 1963 (the US took another twenty years)

Both nations initially pursued Venus with equal fervour, in the hope of uncovering a lush world beneath its clouds. In 1961, Russia embarked on the Venera space programme, named after the Russian word for 'Venus'. Their programme immediately met with challenges: the first probe failed to leave Earth, and the next lost communication shortly after launch. The US faced its own setbacks with *Mariner 1*, but *Mariner 2* successfully flew close to Venus in 1962, revealing the planet's scorching surface temperature and cooling American enthusiasm for further exploration. Consequently, NASA shifted focus to the Moon, leaving Russia to take the lead in exploring Venus.

By 1966, *Venera 3* earned the distinction of being the first human-made object to reach another planet's surface, but it was unable to transmit any data once there, though it was not immediately apparent why. *Venera 4* was more successful, transmitting measurements of the atmosphere as it descended, revealing that Venus' atmosphere is composed of over 96 per cent carbon dioxide, with only trace amounts of nitrogen and other gases. Having descended to 25 km above the surface, it reported that atmospheric pressure was already reaching over twenty times that of Earth. No more was heard as its presumably crushed carcass fell to join the deformed remains of *Venera 3*. A similar fate befell *Venera 5* and *Venera 6* within an hour of entering Venus' atmosphere, but not before they transmitted data revealing atmospheric wind speeds of hundreds of kilometres per hour.

In 1970, *Venera 7* became the first probe to make it through the atmosphere in one piece. At the last moment, though, the main parachute failed, causing it to land hard and topple over. Now pointing the wrong way, the antenna was only able to transmit twenty-three minutes of weak data from the surface before it expired from the heat and pressure. Amongst the information received was a temperature reading directly from the surface: 475 °C, just as radio astronomers had first discovered fourteen years prior. The pressure *Venera 7* measured at surface level was equivalent to that found 1 km below sea level on Earth, enough to instantly force the air from our lungs and the blood from our veins, and crush our bodies. If there had been any lingering optimism regarding life on Venus, it was gone now.

Venera 9 and *Venera 10* finally managed to send back the first images in 1975, using cameras that scanned a sixty-degree field of view to produce images roughly 512 by 512 pixels. One line at a time, bit by bit quite literally, the images were sent back to Earth. This was a frighteningly tense task. The data rate was approximately seventy-two bits per second, so an image would take tens of minutes to transmit, but the *Venera* probes had so far only been capable of withstanding thirty minutes of the intense surface conditions. They did it, just.

The photographs all look very similar, despite the missions landing in different locations on Venus.[6] The probes both landed on hard volcanic plains composed of cooled lava, similar to the basalt found on Earth and the Moon. As lava cools, it can fracture, creating jagged, tablet-like formations. Slabs of rock of all sizes are strewn across the landscape, giving the look of a disused slate quarry. It is not clear what one would choose to explore, even if one could; no landmarks stand out.

We can see to a distance of about a kilometre, but the visibility isn't very clear. The shorter wavelengths of incoming sunlight are scattered in the thick atmosphere, leaving only a faint, yellow haze. On Earth, we see a similar effect during sunset when the Sun's light must travel through more atmosphere due to its low angle. During the day on Earth, the sky is blue because only these shortest wavelengths of light are scattered through the atmosphere and prevented from reaching the surface. At sunset, the longer wavelengths are scattered too, and we are treated to a sky painted with a palette of reds and oranges and pinks.

On Venus, only a few per cent of sunlight manage to penetrate the dense atmosphere and make it to the surface. This meagre light, diffused by the clouds, bathes the planet in a perpetual, overcast twilight. The cloud cover never breaks, and there is not even a hint of the Sun's outline.

By 1975, after the many trials and tribulations faced by the Venera programme, one would think the Russians might have learned enough to build on the most recent successes. But space travel is nothing if not difficult, and while, in 1978, *Venera 11* and *Venera 12* survived the crushing pressure, the ferocious heat, and landed the right way up, in both cases the camera's lens cap didn't release. In 1981, we got the first colour panorama from *Venera 13*, and *Venera 14* managed to release its lens cap and take images too. Unfortunately, the lens cap had fallen directly under the instrument designed to push into the ground and test the surface hardness. One can only assume that by this point, the air on Earth was turning blue with Russian curses.

At last, recognising the hassle of optical photography on the surface, given that each eye-wateringly expensive probe lasted barely an hour, Russia's final Venera missions shifted to radar imaging from orbit. High above the toxic and crushing atmosphere, orbiters from both Russia and the US have used radio waves to map Venus with far greater efficiency. By 1983, NASA's *Pioneer Venus Orbiter* (launched in 1978) had mapped 93 per cent of the planet's surface using radio waves.

The Venera missions that made it through the atmosphere all still lie on the planet's surface, mangled and melted, slowly corroding in the acidic conditions. While travelling to the Moon has been an ambition shared by many nations, Venus is littered with machinery bearing the hammer and sickle insignia of Soviet flags. It is Venus, not Mars, that might properly be called the 'red planet'. To any future alien archaeologists, Venus will look for all the world like a communist outpost pitted against the capitalist Moon.

While Soviet missions were invaluable in providing atmospheric and surface measurements, their most important lesson was that physical exploration is simply not feasible on Venus. Instead, scientists have been motivated to rely on radio reconnaissance ever since.

Radar missions mostly stay safely in orbit around Venus, far from the lethal atmosphere. In shape and size, the spacecraft look rather like the vans used by roaming TV reporters, with the satellite dishes attached to the roof. An orbiting spacecraft's outer structure is largely taken up by solar panels to generate electricity, but it is the lightweight metal dish, usually about 4 m in diameter, that dominates.

These spacecraft use radar to map the surface of the planet they orbit. The colours we choose to represent different altitudes is arbitrary, since radio waves are invisible to our eyes. Scientists working with Venus tend to choose a palette of yellows or oranges to fit with the Venera optical images. Darker shades tend to represent valleys or smooth lowlands, where more of the radar signal is absorbed or scattered, while lighter shades usually mark rough, elevated ground such as mountains or highlands, where radar bounces back strongly.

Early radar maps were low-resolution, blurring together any features smaller than 75 km. At 75-km resolution, satellite images of the Grand Canyon would show just a vague, elongated depression. The detailed features of the canyon, such as its winding course and steep cliffs, would be indiscernible. Yet even at this scale, it was clear that Venus' surface is vastly different from Earth's. Unlike Earth, Venus shows no present-day plate tectonics. Instead, its crust behaves as a single, fragile layer, deforming like the skin on a cooling custard as it is dragged around by the churning mantle below. Despite temporarily puncturing under the pressure of volcanic plumes, the crust never fractured into a lasting jigsaw of continents. The lack of continents on Venus led to a broad lack of geological diversity: volcanic plains cover 80 per cent of Venus' surface, with the rest made up of rugged, inaccessible highlands.

Venus has more volcanos than any other Solar System planet; scientists stopped counting at 85,000 when last studying data from a comprehensive radar survey.[7] Looking at a global radar map, it is striking how the volcanos don't form chains like the Japanese Ring of Fire, or cluster in groups like the volcanic Hawaiian Islands. Without continents, there are no fault lines. Instead, the volcanos are spread evenly across the globe. Unlike the dramatic, towering peaks we know from Earth, Venusian volcanos are deceptively subtle. The slopes are so gentle that we might not even realise we're ascending one, until the detritus of ancient landslides begins to make finding our footing difficult. The high temperatures and slightly lower surface gravity (90 per cent that of Earth) mean lava can flow much further out before it begins to cool down and solidify. As a result, many volcanoes have low, broad profiles, and a distinctive minority are so flat we call them 'pancake volcanos'. The average volcano reaches a modest 1.5 km, roughly the height of Mt Vesuvius, which is an Italian volcano so short in stature even I have managed to climb it.*

What they lack in height, they make up in sprawl. Some are tens of kilometres, even hundreds of kilometres, in diameter. The largest, Maat Mons, has lava flows extending hundreds of kilometres across.

Lava on Venus, flowing at a speed and intensity unseen on Earth, carves deep scars into the planet's surface which show up as darker lines on the radar images. These lava channels are everywhere in the volcanic plains, ending up in deltas reminiscent

---

* Full disclosure, I bussed halfway up it. Still, that you can even take a bus most of the way proves my point.

of the Nile and long-hardened lakes. The longest lava channel known, either on Venus or on any planet in the Solar System, is called Baltis Vallis. It extends a full sixth of the circumference of the planet. If the lava of Mt Vesuvius had enjoyed such reach, its channels would have extended far beyond the nearby towns of Pompeii and Herculaneum to Beijing, Cape Town, New York and New Delhi.

In its distant past, Venus may have been cool enough to harbour shallow oceans, but its geology caused it to develop a hot and toxic atmosphere. On Earth, plate tectonics meant portions of the cooler crust periodically sank into the mantle, while the hot mantle rose up and radiated heat into space, allowing for the gradual cooling of Earth's internal temperature. But Venus lacked tectonic plates, so the planet's crust remained intact, keeping the planet's interior hotter for longer and forcing relentless volcanic eruptions across the globe. These eruptions continuously spewed vast amounts of carbon dioxide into the atmosphere, setting off the runaway greenhouse effect. In addition, the volcanos added sulphur dioxide into the atmosphere, which coalesced into thick, toxic clouds. On Earth, sulphuric acid is removed from the atmosphere in rainfall and is of low enough concentration that it is not usually a health hazard. On Venus, these acid rains do fall, but as they approach the hot surface the rains evaporate, never touching the surface, and replenishing the thick clouds.

Radar mapping has also uncovered Venus' strange craters. One of my favourite science activities to do with young children is 'moon meteorites'. You prepare a flat tray of flour and then sift a thin layer of cocoa powder on top. This represents the lunar rock and the dusty regolith on top respectively. Then, to the children's delight, you let them throw a variety of pebbles and ping pong balls at the tray. As the 'meteorites' strike, white flour erupts from underneath the cocoa powder and fans out in rays.

We can see this pattern extending thousands of kilometres when we look at lunar craters, or Martian craters, or Earth craters. We do not see this with Venusian craters.

Radar close-ups reveal softer craters on Venus, with patterns reminiscent of droplets on dusty land. This happens because incoming asteroids are slowed far more by the thick atmosphere, reducing their impact. The debris is immediately super-heated by Venus' high surface temperatures, causing it to flow like liquid and spread out like a raindrop. The shapes are well preserved, because there is no rain or surface wind to erode them, or plant or animal life to cover them, like on Earth. On the Moon and on Mars, even, craters find themselves eroded by other impacts, as their atmospheres allow a constant barrage of meteorites to get through. Venus' atmosphere prevents most small asteroids from getting through at all and, though some larger ones do penetrate it, the number is small. Even taking the protective effect of the atmosphere into account, the number of craters is low, which has led astronomers to estimate that the surface of Venus is only a few hundred million years old. Intense volcanic action in the past has constantly resurfaced the planet, with cooling lava like tarmac smoothed over potholes.

To visit all the sights worth seeing, one must hitch more than one ride with a radio wave, back and forth between satellite and surface. Let's leave the volcanic plains that cover most of the surface, zip back up to our orbiter, and head down for one more journey, this time to the highlands. Because we can direct our beams of radio waves where we like, we can achieve this change of location, equivalent to travelling thousands of kilometres on the surface, in several milliseconds.

The Venusian highlands make up about 10 per cent of the surface, with two main regions standing out: Ishtar Terra and Aphrodite Terra. Ishtar Terra, roughly the size of Australia, features Lakshmi Planum, a volcanic plateau 2,500 km across and 3 km high.* The towering Maxwell Montes, with peaks reaching above 10 km, taller than the Himalayas, dominates this region.† Aphrodite Terra, which stretches over a third of Venus' circumference, includes Diana Chasma: a deep chasm that only radar can map through the clouds. These measurements show that the chasm descends 2 km below the average radius and 4 km below the surrounding ridges, making it the lowest point on the planet's surface.

The most recent space-orbiter radar maps date from 1990–4 during the NASA *Magellan* mission. They achieved a resolution down to 100 m, making radar maps of Venus so comprehensive that they have been rendered in 3D, such that we can really imagine ourselves exploring the surface.[8] The images have the feel of a 1990s first-person shooter, with the same distortions one notices when exploring Google Street View. But the detail is enough that you feel like you are really exploring all the features that make the surface of Venus one of the most unique places in the Solar System. These images are frozen in time, though. They are decades old, from missions that have long since been decommissioned and thrown themselves out of orbit and into a fiery and fatal descent towards Venus. Every new orbiter mission in the Pioneer, Venera, Vega and Magellan series deepened our understanding of Venus, but, as all-encompassing as they were in terms of mapping, they were still, like the six optical images from *Venera 9* and *Venera 10*, only brief snapshots in time. Our next close encounter with Venus is planned for the early 2030s, at best.

★

* This means it is 3 km 'above the average radius'. Venus lacks a sea level, so the planet's topography is measured relative to an average planetary radius of 6,051.2 km.

† This is the only feature on Venus named after a man, James Clerk Maxwell, the father of electromagnetism, just before the International Astronomical Union made the rule that all things Venus must be named after women.

We've gathered an impressive dossier on Venus' troubled past. Despite being one of the most secretive and hostile targets in the Solar System, Venus has become a testing ground for theories about planetary formation and habitability. This is not just about completing the family history of our Solar System; it's about expanding our understanding of where life might flourish, or perish, beyond Earth.

Venus was believed to be such a viable candidate for life that it broke humanity's heart when its surface turned out to be barren. This revelation forced us to rethink our assumptions about what makes a planet habitable. We've learned that proximity to the Sun isn't the only clue in this cosmic investigation. It's increasingly clear that tectonic activity might be key to sustaining life. The movement of the continents helps to regulate a planet's climate by releasing pent-up energy and ensuring carbon is distributed between the oceans, atmosphere and interior. Their interactions on Earth have created diverse landforms and environments, which foster biodiversity and drive evolution, while Venus was left to snap, crackle and pop, its lava oozing across enormous plains.

It is easy to come up with convincing theories of tectonic processes, but we still need to test whether these predictions are correct: are the landslides due to seismic activity? Is Venus volcanically active right now? By repeating radar measurements over time, we can see whether individual rocks have moved due to the rumbling of the ground below, observe any new volcanos which erupted and assess whether there have been new lava flows covering old craters.

Some answers take time, and in this game of cosmic espionage, patience is as valuable as cutting-edge technology. Ground-based radio astronomy enables the persistence needed for long-term surveillance, unlike the brief visits of space probes. For example, only through fifteen years of patient repeat radar measurements were we able to ascertain that Venus does indeed have a variable day, and that the atmosphere has the power to speed up or slow down the planet's rotation.

While the *Pioneer* probes produced exquisite maps by flying to Venus, astronomers on the ground at the Arecibo Observatory were able to *match the resolution* of those space probes only one decade later, in 1988. By that time, radio astronomy was a well-established field and all sorts of improvements had been made to the technology. I cannot underline enough what an impressive feat this was: mapping the surface of a planet 200 million km away, down to a 2-km detail. If we were to give your phone the same ability, you would be able to take a selfie and then zoom in to see the bacteria on your skin. Because our radar 'zoom' function is the best there is, we can take close-up photos from another planet. Our comfortable, hospitable, habitable planet.

In our efforts to peer beneath Venus' impenetrable veil, the tools of radio astronomy remain our most reliable assets. They allow us to gather intelligence on a planet that resists every attempt to reveal its secrets, reminding us that sometimes the most powerful agents of space exploration are a team of radio waves that can observe, wait and infiltrate when the moment is right.

5

MARS

> If you can't take Mars for what she is, you might as well go back to Earth.
>
> Ray Bradbury, *The Martian Chronicles*

Unless you're one of the few astronauts aboard the International Space Station, it's likely you're reading this book from Planet Earth. Despite the optimism of the 1960s, which predicted moon-bases and cities on Mars by the dawn of the millennium, these celestial bodies remain uninhabited by humanity. After all, why spend trillions of dollars attempting to survive elsewhere, when our home planet offers everything we need?

For thousands of years, humans have spread across the globe, transforming environments to suit our needs. We've dammed rivers, cleared forests, built outwards and upwards and exploited every available natural resource to power our ambitions along the way. Earth has been good to us, even if we have not been so considerate in return.

In stark contrast, as we've seen in previous chapters, the Moon, Mercury and Venus offer no such hospitality. The Moon is airless, devoid of protection from solar radiation. Mercury, while it harbours some frozen water, swings between extreme heat and bitter cold. Venus, once thought to host life, is a hellscape where even the toughest robotic explorers perish within an hour. We've sent probes to study these worlds to satisfy scientific curiosity, and men to the Moon to assert political dominance, but none of these missions sought new homes for humanity.

And yet, it cannot have escaped anyone's notice that things down here on Planet Earth are getting a little dicey. In building our own industrial, economic and technological paradise, we have systematically dismantled the natural environment that nurtured our evolution. As climate change accelerates, sea levels rise and forests burn, suddenly these alien worlds don't seem quite so inhospitable.

Well, okay, Mercury and Venus are never going to make it onto our property wish list, but the Moon? Could we make it work as temporary accommodation, a stop gap while we look for something bigger? We stare at floorplans and glance at each other doubtfully as we work out where everyone would sleep.

Then there's Mars. A small, red planet, just down the road. It's been on the market for a while, and the listing comes with a warning: it's cold and the roof leaks solar radiation. It's half the size of our current home and has no garden. Still, at around 225 million km from Earth, it is within reach, and a manageable commute from the Moon. It's a real fixer-upper . . . but hey, we're getting desperate. With Earth's growing crises, moving to Mars is no longer the distant dream it once was. If the new space race continues to progress at the speed we have seen so far this decade, the first person to die on Mars has likely already been born. The first person to give birth on Mars might already be alive, ignorant of their destiny as a parent of the first bona fide Martian.

Before we pack our bags, we need to know what we're getting into – and that's where our study of other planets comes in handy. Mars is easy to see by eye, and with an amateur telescope one can make out the red hue of the iron oxide surface and its contrasting white polar ice caps. But anyone who has bought property knows it pays to really do the research, to actually visit the property, look for cracks in the walls and sniff out damp. Luckily, we have explored enough nightmare properties with our radio telescopes so far that we know the warning signs.

There are no shortages of places to settle on Mars. It is empty, after all. But different regions of the red and dusty planet offer different advantages . . . and disadvantages. And I'm not talking about heated pools and air conditioning. Here, the most desirable plots have two basic things: access to water and shelter from the solar wind.

First on our checklist is water. It's a non-negotiable demand to make of any cosmic estate agent. Water is life's fuel – without it our bodies begin to shut down after only three days. Each of us needs litres a day, yet Mars is a dry planet. The wide riverbeds and deltas that scar the planet tease us with the geological memory of a time when water flowed freely, if irregularly. It is unlikely that there were ever oceans or rivers as we know them: stable bodies of water that have existed for hundreds of millions of years. Instead, when Mars was younger, most of its water appears to have been locked underground, frozen in permafrost or trapped within porous rock. Bursts of volcanic activity heated these subterranean stores, releasing catastrophic flash floods that carved through the Martian surface like a knife through butter. But the water didn't linger; it quickly seeped back into the ground or sublimated into the thin atmosphere. So, while water flowed on Mars around 4 billion years ago, it did so briefly and intermittently. Its rivers and lakes were transient, driven by an underground water cycle that sputtered rather than surged.

Mars missions over the past two decades have revealed a planet rich in clues about water. The good news is that the ingredients of water – hydrogen and oxygen – were found in quantities that suggested each cubic foot of Martian soil could yield up to two pints of H_2O. Now for the bad news. Martian soil was also shown to contain perchlorate, a dangerous chemical, and enough to be downright toxic to any human. Strolling on the surface of Mars is no walk in the park. While the sky irradiates you, the soil poisons you, all the while taunting you with the intangible water right beneath your feet.

As well as soaking into the soil itself, a lot of the liquid water froze rapidly in the Martian night and Martian winter, locking itself away in huge ice deposits. Over time, the dust of the Martian soil blew a thick blanket over all but the greatest ice deposits in the polar regions.

Luckily, we have radar instruments that can probe below the surface to help us investigate the underground deposits left over from ancient floods. These layers of ice can lie kilometres below ground, completely obscured by the red dust that envelops the sub-polar regions, but this is a depth we can just about reach with radar. As we saw on Mercury, ice reflects radio waves in a different manner from dust, and even different kinds of ice reflect differently so, like a high-tech dowsing rod, radar can guide humanity's explorers to water.

But do we really need this technology when we can see the white of the ice caps on Mars with our own eyes? The obvious place to settle might be these polar regions. There is more than enough water there. It is estimated that, were we to melt the ice caps, the entire surface of Mars would be covered by a global sea

20–30 m deep, enough to satisfy the thirst of trillions of humans over their lifetimes. But these humans would need to really like the cold. On Earth, in Antarctica to be precise, the coldest temperature recorded is about −88 °C. On Mars, the temperature at the poles plummets to −125 °C during the winter. Even getting machinery to work at these temperatures would be a problem. Batteries, for example, depend heavily on the ambient temperature. The charge on your phone might last a whole day while wandering around London or New York, but it would only last minutes at the North Pole. If it worked at all.

The temperatures elsewhere on Mars are more manageable, though they're never going to be entirely comfortable; the planet has little atmosphere to retain the heat from the Sun, which is further away to start with. At the equator, you might enjoy a 20-°C day in your (dusty) garden for a few hours during the summer, but at night you'll still find yourself confronted with temperatures down to −70 °C. And in the winter, you can knock both daytime and nighttime temperatures down another 30 °C.

Since a neighbourhood near the icy poles is not going to work, we need to find a different source of water. One of Mars' massive ice deposits can be found in a region called Utopia Planitia, which hosts as much water as Lake Superior in a sheet around 100 m thick. 'Utopia' would look good on the Change of Address cards, too, though it is a touch misleading. Utopia Planitia is a plain within the larger region of Utopia, which has the questionable honour of being the largest impact crater in the Solar System. With the crater measuring 3,300 km across, scientists have estimated that the asteroid was 600–700 km across. The crater is so huge that if you were to stand at the centre of it, you would not realise you were within one at all. It is flat and smooth as far as the eye can see, hence the name 'Utopia' – an untouched, ideal place.

But herein lies the problem with this plot of land. The lack of features and contours leaves us unsheltered, entirely exposed to the elements. Our homestead would be isolated, and vulnerable to attack from above.

Mars lacks a magnetic field, and so it cannot shield itself against the onslaught of high-energy particles hurled from the Sun. Martian surface-dwellers will bathe in solar radiation hundreds of times stronger than that received on Earth, enough to cause major genetic mutations and cancers. And if one is willing to take that risk to wake up with a Martian sunrise, even the bravest will have to run for shelter during the worst solar storms. Then, the sudden peaks in radiation will be enough to cause acute, and potentially fatal, radiation sickness. Utopia is not looking like such an attractive proposition.

The third potential neighbourhood is on the opposite side of the planet, nearer the equator. This is Valles Marineris, one of the most spectacular sights in the Solar System. It is the largest canyon system on Mars, its length equalling the width of the United States. It makes our Grand Canyon look like a crack in dried mud. Proximity to the equator makes it naturally warmer here, but our interest is less in the daytime temperatures than in how these vast, deep walls might provide some natural protection from two of the biggest nuisance factors: temperature variations and radiation. Settling at the bottom of the ravine, we can enjoy a slice of Martian sky, and a smaller dose of radiation, while benefitting from the slightly warmer air trapped between the walls. No property is perfect, however, and while this one has climate control and sun protection, we can't be completely sure there will be water ice on tap. There may well be some buried deep within the walls, but is it worth the risk of moving there without a back-up plan? Perhaps a place to keep an eye on for when we've got a solid base somewhere else, and are looking for a second home.

So, what do we have left on our property shortlist?

Radar's strength as an estate agent is in discovering the hidden gem, in looking past the grime and the bad décor, to the solid walls and foundations that will make a good, stable home for generations. Humans are, by nature, surface-dwellers, but Mars is not our natural habitat, and our Earth-centric preference for topside living might just have to go. We have seen already how radar telescopes on Earth are able to find vast lava caves on the Moon, produced when ancient magma flowing through underground channels cooled, shrank and solidified. Radar has found lava tubes on Mars to be just as ubiquitous, and the protection they can offer is even more important.[1] One can imagine the Moon as somewhere the cosmic commuter might stay for a few months while waiting for a shuttle to Mars, or maybe a few years while the industries required to sustain interplanetary travel are built up. But humans will spend whole lifetimes seeking shelter from the radiation on Mars.

We've knocked on the lava-tube walls to check they are sound, now to investigate the plumbing. And this is where radio astronomy has really found the deal-clincher, because radar maps have revealed not only where lava tubes and ice sheets coexist, but also signs of subsurface water.[2] Under normal circumstances, it would be far too cold for liquid water to survive on Mars, and so if these underground lakes do exist they will likely be either very salty (significantly lowering the freezing temperature) or warmed by even deeper volcanic activity. Who doesn't want a house with both running water and underfloor heating?

Once radar has found us a new home, one can imagine a whole Martian civilisation growing up underground, shielded by rock, chipping away at their ice supply. Elders will tell stories about a planet on whose surface they walked and lived, making mudpies and bathing in the sunlight. They will be met with incredulous young eyes. Perhaps, lacking any seaside to speak of, family day trips will be made to the surface: a rare outing, where the radiation toll is counted and accepted. At the end of a tiring day exploring the local Martian terrain, the family will return to the lift, and the kids will strain upwards on their tiptoes to glimpse the last flash of red soil as they sink once again to their subterranean shelter.

One thing is certain, no matter how long the daytrip, there is little chance of bumping into the locals. Normally, the first thing you do when you move house is get to know who lives next-door. But it looks like we will have Mars largely to ourselves. Despite decades of searching, and several false starts, we have failed to find definitive signs of life on Mars. We know enough about its volcanic and geological history to know that, once, it was warm enough and wet enough to host simple, microbial life, and so it is quite possible we are simply looking in the wrong place.

The first landers we sent to search for direct signs of this life, *Viking 1* and *Viking 2*, dug into the soil where they landed and sifted and searched for the chemical signatures, but to no avail. With hindsight, clawing randomly through the dry and irradiated dust and looking for life is rather like sifting through the Sahara and being disappointed we found no evidence of fish. Martian life, we can assume, would need water just as much as we do, and so the place to look for life, current or past, is where the water is. Portable radar units on the modern Mars rovers now aid them in identifying the parts of the surface that host subterranean ice, such that they can search for life where it is most likely to thrive. Perhaps, as we seek a source of water, we might stumble across some new neighbours, though they will be far from evolved enough to pop round for tea and a slice of cake.

Let's imagine that radar has helped us choose the perfect Martian neighbourhood. Now, how to get there. It will be quite the move, so planning is everything. As we saw with the inner planets, radio waves can measure the distance, rotation and orbital speed of nearby celestial bodies. Without this data, travelling to

Mars would be like throwing a dart at a moving dartboard, blindfolded.

Timing is everything, because Earth and Mars are moving relative to each other, at different speeds and along wildly different orbits. While Earth is travelling in an almost circular orbit around the Sun, Mars takes a more elongated route, with a Mars-year being 687 days long. At their greatest separation, Earth and Mars are 400 million km apart, on opposite sides of the Sun. This immense distance is effectively impossible to traverse efficiently with current technology due to the cost of lifting so much fuel out of orbit. Luckily, once every twenty-six months, Mars and Earth come close enough that they are only 54 million km apart. The removal van, our spaceship, will only take bookings once every other year, then, and the journey will still likely take six to nine months.

When we set off from Earth, we need to make sure we are aiming not for where Mars is now, but where it will be by the time we get there. Orbital dynamics are nothing new and astronomers have been able to predict when and where Mars would appear in the sky for thousands of years. Mars is so far away, though, that tiny errors in the trajectory will be massively exacerbated over the long journey. If a ferry crossing the River Thames heads one degree off course, it will land a few metres from where we had intended – not too much of a problem, and easy to correct. If we were to launch a spaceship from Earth with that kind of an error, even when Mars was at its closest, it would miss Mars by almost a million kilometres – that's over twice the distance between the Earth and the Moon.

To be sure that our space probe arrives at the right planet, our aim has to be accurate to within one ten-thousandth of a degree. For reference, the angular separation of the minute markers on a clock is six degrees. This level of precision requires knowing both the position and velocity of Mars to an astonishing level of accuracy. What's more, because Mars is also rotating on its axis, we need to know that speed too, so we time our journey to arrive on the right part of the planet.

The first thing to do is send a strong burst of radio waves towards Mars and wait. We stare at the clock for seconds, a minute, ten minutes . . . until suddenly the numbers freeze in reaction to a returning burst of radio waves. An echo that has reached our antenna. The time that has elapsed tells us exactly how far away Mars is. Radar is the most accurate way of determining distances to nearby planets, short of travelling there and using some kind of fancy NASA space-tape-measure along the way. We can take these radar measurements at various points in our respective orbits, measuring the distance and speed of Mars to the accuracy we need, to be certain that we arrive close to the survival supplies sent in advance. Get lost in the wrong neighbourhood on Mars, and you could easily starve or run out of oxygen before you find your way home.

We will be entirely reliant on oxygen tanks when walking outside on Mars, as the air mostly consists of carbon dioxide. It's not just that the air doesn't contain enough oxygen for it to be breathable, it is that there isn't much air overall: the average air pressure on Mars is around 0.6 per cent of Earth's atmospheric pressure at sea level. Mars is also known for its dust storms, which can last for weeks or months, blocking sunlight and rendering solar panels useless. In addition, they alter temperatures dramatically and cause significant drops in visibility, so we really don't want to be out wandering for long.

We can learn a lot about Mars' atmosphere without ever setting foot on the ground, thanks to a radio astronomy technique called 'occultation'. This method involves instructing a satellite that is orbiting Mars (or any planet or moon we can reach) to send a continuous stream of radio data. It doesn't even really matter what: it could equally be data describing Gustav Hölst's Mars suite or Taylor Swift's back catalogue, or both, as long as it is on repeat. As the satellite moves behind the planet, for a brief time the signals pass through the alien atmosphere on their way to Earth. The signals are then blocked entirely, before there is a similar brief moment as the satellite peeks out on the other side. The signals are bent and scattered and delayed by the Martian atmosphere, and the contorted data we receive enables scientists to model the composition temperature, pressure and density of the atmosphere at various altitudes. This method, rather like launching a thousand messages-in-bottles and charting ocean currents from where each one comes to shore, has been employed by missions such as *Mars Express* and *Mars Reconnaissance Orbiter*. Radio occultation data can also help us track more transient atmospheric issues that might affect missions, like the formation of dust storms.

During the six-month journey to Mars, we will steadily watch Earth become just one bright light amongst many in the sky. Meanwhile, in the forward direction, we will see the red planet get bigger and bigger. There will come a halfway point when our eyes see both Earth and Mars as mere star-like points and we will feel truly untethered. The radio antenna on our ship will be our only link to the planet we left behind, sending our updates and questions through the dark to the waiting metal ears of the Deep Space Network of radio telescopes on Earth. At first, we will be able to video call while looking at the big blue globe out of the window, and it will feel almost like we're speaking to our friends from down the road. Then, weeks in, the increased distance between Earth and the antenna on our ship will mean our signal is not so strong, and we might be reduced to phone calls. By halfway to Mars, even the radio light takes ninety seconds to get to Earth. That might not sound like a long time, but bear in mind how annoying it is when there is just a two-second lag on a terrestrial video call. The pace of conversations will reduce from that of a volley of WhatsApp messages to a stilted email exchange, reminiscent of a time when patience was part of keeping in touch.

Nothing can be done about this delay, even when we are settled on the red planet and have developed scientific industry to rival Earth's. We cannot speed up light, and we cannot move Mars closer to Earth. We can look forward to our closest approach every twenty-six Earth-months, when Earth will shine brightly, albeit no more brightly than the brightest star in Earth's night sky. Then, messages will take a mere three minutes to reach Earth. At largest separation, however, we will squint into the night sky and struggle to pick out Earth from among the stars, knowing that any messages we send won't be heard for more than twenty minutes. Still, even with the delays, radio communication will undoubtedly be a psychological lifeline for the first generation of Earth emigrants.

There are moments in life when you grow up in an instant. The safety of your worldview is shattered by a personal or global event and you begin to understand something that hitherto you had not really thought much about. These moments happen in science, too. There are moments when a hope becomes a reality and then a normality: the detection of gravitational waves for example, or the imaging of a black hole. Then there are the moments when hope is turned sour, and you realise the consequences of taking a risk. When it comes to caution in science, I grew up in an instant on Christmas Day 2003, though I don't think I appreciated how it had affected my later career decisions until over a decade later.

I was a teenager with an interest in physics and, though space was not my main interest, I couldn't help but follow the European Space Agency (ESA) *Beagle 2* mission as it made its way to Mars. It was all over the news because it was a rare, UK-led mission, with a colourful character of a mission leader whose enthusiasm exuded from the TV screen. He was so excited! How could I not be, too? Named after Charles Darwin's ship, *Beagle 2* would land on Mars and deploy instruments designed to search for life. For a British kid like me, in a country where space exploration received very little investment, this sudden eruption of excitement from the UK population was infectious.

When the *Beagle 2* lander mission was successfully deployed from Mars' orbit in December 2003, everyone held their breath, waiting for *Beagle 2* to send a radio communication indicating it had landed safely on 25 December – Christmas Day. But the message declaring success never came. In fact, there was not even a message suggesting any form of failure. *Beagle 2* launched from its orbiter into the high seas of the Martian atmosphere . . . and simply disappeared. Back then, the satellite cameras could not reach the resolution needed to search for the tiny lander, so there were no images of a crushed machine, and no closure. The image broadcast all over the news was instead that of the mission leader Colin Pillinger, looking utterly desolate in the control room. They had done everything right, everything had been well planned and executed. But it didn't matter. All that enthusiasm and hope and excitement, all that planning and time, all those tens of millions of pounds, gone. Maybe NASA can swallow these disappointments. As we saw on Venus, the Soviet Union certainly could. But the UK? No, this failure would set us back for a long time.

For some reason, that image of a defeated scientist, drained of the life energy I had seen during the build-up to launch, stuck in my mind, and it pops up with every mission failure I am exposed to in my scientific career. I think that, when it came time to choose my allegiances, this image really added weight to my decision to become a researcher in radio astronomy, and not optical. By the end of my PhD, I was very aware that the radio telescopes I had worked with were tangible, reachable by plane or train, and fixable. They were made of parts that one could buy in a hardware shop, and cheap parts at that. The idea of working on something for years, decades even, all the time knowing the risk of a complete mission failure was close to 50 per cent, was not something I could stomach. I simply couldn't face having my heart broken like the astronomers behind *Beagle 2*.

The team pronounced *Beagle 2* dead in February 2004, but real closure only came in 2015, when a new Mars orbiter found the *Beagle* lander in its upgraded satellite imagery.[3] Surprisingly, it had landed perfectly, but it stood like an inert statue in the Martian dust. From the images, it was clear that the solar array and communications antenna had not fully deployed, robbing *Beagle 2* of both the means to generate energy, and the ability to cry for help. Without the ability to receive instructions, any spacecraft is rendered redundant. I wonder if it might be the same for the first humans on Mars. Without instruction, without a connection to the planet they were born on, will they flounder?

When I imagine the first Mars settlement, I don't imagine a futuristic space base. Instead, I think Mars will look much as it did before we moved in: desolate. The humans will be siloed beneath the red earth, a mix of nationalities struggling to mesh their cultures on a planet where these divisions have no context. When *Apollo 11* landed, one of the first things they did was plant a flag, and perhaps it will be the same on Mars. But I think the first visitors will also be very quick to place down the simple, sleek, metal frame of an antenna. The houses and offices may all be hidden in lava tubes, robbing us of a Martian city skyline, but above ground there will be the outline of a radio telescope, built to maintain a connection with those we left behind.

For now, Earth remains our home, and most of us are not ready to leave just yet. But like tenants watching the cracks widen in the walls of a crumbling home, humanity would be foolish not to start planning a move. Soon, we might have no choice but to find a new place to live. Thanks to radio astronomy and radar, we've mapped out Mars' hidden corners, found water buried deep beneath the dust, and calculated the safest routes for the journey ahead. These long wavelengths of light will keep us grounded on Earth even as we step out towards new horizons, ready to turn this red and dusty planet into somewhere we can call home. It's not perfect, but with the right tools, it just might be the fresh start we need.

6

THE ASTEROID BELT

'If the Earth gets hit by an asteroid, it's game over. It's control-alt-delete for civilisation.'

Bill Nye, 2019

You never forget your first glimpse of Saturn through a telescope. Until that moment, it's likely you have looked straight at it without realising: it looks like just another bright star wandering through the constellations. But, through a telescope, it becomes unmistakeable: a perfect sphere encircled by a luminous belt of light. Saturn's rings are far from unique in our Solar System. Jupiter, Uranus and Neptune all have ring structures too, but theirs are ghostly wisps in comparison, barely visible even with the most powerful telescopes. Saturn's rings, in contrast, are bold and brilliant. For many an astronomer, amateur or professional, seeing those famous rings for oneself, with one's own eyes, marks the moment at which we gave our heart to the stars and vowed a lifelong devotion to the night sky.

My first experience of seeing Saturn's rings came after I bought my first telescope at thirty-three years old. How does an astronomer get to thirty-three without owning a telescope, you ask? The short answer is: the internet. I never had a telescope growing up and couldn't afford one when I was studying astronomy at university. Then, by the time I was a paid astronomer with access to images taken by the best space telescopes out there, I lost interest in personally seeing Saturn through a small telescope in my back garden. What could compete with those high-resolution images, so crisp and detailed, so perfect that they almost looked artificial? Besides, what barista wants to go home and spend their evening making coffee?

Gradually, though, I formed an appreciation for the craft of astronomy, as opposed to just the study of it. I wanted to tinker, to point, to search, not just flick through photo albums somebody else had had the joy of creating. I came to love the ritual of setting up a telescope and the meditative silence of the thirty minutes when I would just sit outside, allowing my eyes to adjust to the dark before observing. Or perhaps I just loved the thirty minutes of peace straight after putting the kids to bed. Either way, I was thirty-three, a seasoned astrophysicist, and my first telescope had arrived.

For the first few clear nights I was captivated by the Moon and couldn't look away. Then, I decided to try my luck with the planets. I spent an hour struggling to calibrate the telescope to point where I wanted it to, a skill we never learn as professionals since our telescopes are always automated. Then, sweeping the sky in the direction of Saturn, a bright fuzzball suddenly swung into view. With a careful turn of the focus, the outline sharpened, revealing Saturn. It could be nothing else. I gasped out loud, frozen in place for fear I would nudge the telescope and lose this incredible image. I stood transfixed for some time, until eventually I regained the ability to move, grabbed my eight-year-old daughter from inside, stood her on a box, and asked her to look. Her reaction was the same: her eyes welled up with pure awe.

Composed almost exclusively of pure water ice, those signature rings are dynamic, shifting structures, and extraordinarily flat – on average just tens of metres thick. But while they seem to hover serenely around the planet, a history of violence lies beneath that peaceful facade. The most likely explanation of their origin is that they formed from the debris of two icy moons that collided in a cataclysmic event, or from a single moon spiralling too close to Saturn, torn apart by gravitational forces. Either way, the trail of rubble was captured by Saturn's gravity. Over time, the accumulation of swirling debris spread and flattened into a disc, like dough being spun into a pizza. The rings of Saturn are not solid, decorative bands, but an ice-cold memory of destruction.

As we have seen on Mercury and Mars, ice is a highly reflective material when it comes to radar. But Saturn is about fifteen times further than Mercury . . . surely too far for radar? Well, not quite. An image of Saturn using radar is nearly as awe-inspiring as an optical one, for me, though for the audacity of the exercise rather than the beauty of the product.[1] We have attempted the feat only once, between 1999 and 2003 using the Arecibo 305-m dish. The result was, and to the best of my knowledge still is, the furthest radar image ever to echo back to Earth, right at the limit of the technology.

They are striking images, because the rings take even more of a central role, while the planet itself is jarringly absent. Saturn itself is radar-dark, invisible against the bright, reflective halo of its rings. The gas giants such as Jupiter and Saturn are, as the name suggests, made predominantly of gas. Radio waves do not rebound in the same way from a gaseous atmosphere as from a solid surface (a benefit when we want to see through the atmosphere of Venus or Earth).

The huge distance to Saturn (on average 1.2 billion km) results in the radar signal weakening dramatically by the time it echoes back from the rings. A radio light signal returning to Earth from Saturn will be 100 trillion times weaker than the same signal echoing from the Moon. So while the icy particles of Saturn's rings do reflect the signal, the returning image is grainy and poor-quality.

These images of an absent planet at the centre of some blurry rings is crude and, if I'm brutally honest, of limited use scientifically. But, my word, they were taken in a jungle in Puerto Rico, right here on Earth. That we can design a telescope capable of sending radio waves to Saturn, and then detect the faint whisper of a signal that returns two and a half hours later, is nothing short of astonishing. In many ways, this blurry image feels more monumental to me than the crisp optical close-ups returned by the space probes that have actually visited Saturn.

Saturn marks the limit of planetary radar as we know it today, any further and the echoes simply become too faint to detect. As captivating as it is to look at Saturn's rings with radio waves, we need to turn our attention closer to home, where planetary radar is less of a novelty and more critical to our survival.

Between Mars and Jupiter lies a far less pretty and infinitely more perilous ring of debris: the asteroid belt.

The asteroid belt marks a boundary in our Solar System, a line drawn with stellar dust by Jupiter. The rocky planets lie on one side: Mars, Earth, Venus and Mercury. Meanwhile, on the other side lies an icy realm, dominated by the gas giants, Jupiter and Saturn, and the ice giants, Uranus and Neptune. This stark difference came about because of the sharp drop in temperature in the young Solar System: heavier rocky material condensed closer to the Sun, while the lighter gases could only form further out. In addition, light elements that are left over, swirling around a newborn star, are soon blown to the outer regions by the solar wind.

As the disc of material rotates around the young star, small granules of rock or ice (depending on where you are in the Solar System) collide and stick together, slowly growing. Like a snowball rolling down a hill, these pebbles orbit the Sun, collecting more material and becoming larger. The more mass these bodies accumulate, the stronger their gravity becomes. Eventually, they don't just adhere to material they touch, they pull in debris from surrounding space, clearing a channel through the stellar disc like a snowplough. In the outer regions, beyond the asteroid belt, compounds like water and ammonia freeze into icy grains, which stick together more easily than rock. This allowed the outer planets to grow rapidly, drawing in the abundant gases that formed their vast, low-density atmospheres.

Jupiter won the race for mass in our Solar System. It began forming slightly earlier than the other planets and so got a head-start, gathering two and a half times more material than all the other Solar System planets combined. As it ploughed through the disc, any leftover building materials within gravitational reach were pulled about with such force that they couldn't come together to form their own new planet. These pebbles and rocks and grains of dust were left to form a vast ring of rubble orbiting the Sun. If you think big enough, the Sun has its rings just as Saturn does. But, while Saturn's rings are a matter more of curiosity than of concern, the asteroid belt contains rocky fragments that could be lethal to all life on Earth. Through collisions among themselves, or gravitational nudges from Jupiter as it loops around the Sun, asteroids can be knocked from the belt and potentially set on a collision course with Earth.

In Roman mythology, Jupiter was the supreme ruler of the gods, known for hurling thunderbolts at his enemies. The planet

Jupiter has been launching space debris for billions of years, and our Solar System is full of these asteroids, each leaving a trail of rubble behind it. When Earth passes through these trails, the incoming meteoroids compress and heat the atmosphere around them, creating a bolt of light in the sky – a shooting star, or, more scientifically, a *meteor*.* The Geminids, for example, which you can watch out for in December, are left over from the 6-km-wide asteroid 3200 Phaethon, which orbits the Sun every eighteen months, regularly replenishing the supply. These rocks spread out into a wide trail, through which Earth can plough once again.

Earth will not cross paths directly with 3200 Phaethon for the foreseeable future, if ever, because its orbit is tilted such that it rarely intersects with Earth's own orbit. Even when the timing does work out, 3200 Phaethon is still millions of kilometres away even at closest approach. So that's one less rock to worry about. But Jupiter has already catapulted vast numbers of rocks towards us, many large enough to cause city-wide, or even planet-wide, devastation . . . and there is no shortage of ammunition left in the belt.

There are so many asteroids, all travelling at high speed, that trying to count them is like counting ants as they scuttle around an anthill. There are about 200 known asteroids above 100 km in diameter. Ceres, the largest asteroid we know of, is about 940 km and could cover both London and Berlin. Smaller asteroids can be the size of a city block, or a small car, and there are millions, probably hundreds of millions, of them hurtling through space. In addition, there are vast numbers of smaller bits of rubble – *meteoroids* – which can be as small as a grain of sand. Altogether,

* The naming conventions seem designed to confuse, but, as a rule of thumb, meteoroids are small enough that they cause minimal, if any, damage, while asteroids are large enough that you'll be too busy evacuating the impact zone to enjoy the fireworks. Mete*ors* are the pretty light show, and if any part of the meteor*oid* survives the burning trip through the atmosphere and lands on Earth, that rock is then called a meteor*ite*. If an asteroid makes it all the way through and hits Earth, we don't call it a meteorite. We call it a disaster.

the total mass of the asteroid belt is only 4 per cent the mass of our Moon, with that single asteroid, Ceres, constituting almost a third of this mass. Still, one only needs to call to mind the mass extinction of the dinosaurs to understand how much destruction just one 10-km asteroid can wreak.

We are under constant attack from this detritus of planet formation, hurled towards us by a giant planet pacing the borders of his kingdom. Earth would only have to be in the wrong place at the wrong time to collide with any one of these threats, and our only hope lies in early detection and pre-emptive action. In this context, planetary radar isn't a diverting scientific curiosity, it's a tool of survival, capable of tracking and predicting the movements of these rogue rocks.

Radar can help us predict the speed and direction of an asteroid's movement in exquisite detail, and we can chart an asteroid's path to easily predict to the day, even to the hour, when a collision will happen. The more notice we have, the better we will be prepared to survive it. The trouble is, we don't know where all the asteroids are, especially if they are on large orbits and have not been close to Earth while we've had the technology to see them. We have to keep a constant watch.

Much of astronomy involves a passive engagement, patiently waiting and ensuring we are observing the cosmos from the correct vantage point at precisely the right moment. We are not privy to the exact timing of a supernova explosion or the intensity of an approaching gravitational wave; we can only use our data to make predictions. In contrast, the field of 'planetary radar' allows us to actively interact with the vastness of space. With radar, we choose the signal, its form, timing and direction, and wait for the echoes. It represents an *active* approach to astronomy. Radar is a form of communication. In the case of planets, it is to say 'Hi! What are you made of?' In the case of asteroids, it is to say 'Hi! Are you just passing by or here to kill us?' There are plenty of asteroids to talk to – there are over *one million* known asteroids within the Solar System. All we have to do is engineer the right technology to reach them.

Optical telescopes take their turn to guard Earth only at nighttime, in clear skies. In war, though, one has to be prepared for an assault at any time. Radio telescopes are our sentinels, quietly scanning the skies for any signs of an approaching assault, day or night, and in any weather. They are the early-warning system that protects our terrestrial borders just as Home Front radar systems guarded our coastlines during the Second World War.

Radio astronomy has long been integral to asteroid detection, particularly in the UK. During the Second World War, Bernard Lovell, a physicist at the University of Manchester, was hired to work on adapting radar for airborne use, enabling pilots to navigate through darkness and target submarines and cities through cloud cover with unprecedented precision. As with other radio astronomy pioneers, including Karl Jansky with his 'cosmic hiss', and Stanley Hey with his radar blackouts, Lovell became a radio astronomer by accident, in an attempt to eliminate persistent interference plaguing his radio equipment.

Most of the radio waves received by Lovell's airborne radar systems were the desired echoes of signals sent by his own antenna, rebounding from target planes or submarines. On top of these, though, was a sprinkling of radio signals he could not explain. He wondered if he was hearing reflections from the trails left by high-energy particles called cosmic rays entering Earth's atmosphere. These cosmic rays are expelled at almost the speed of light from various sources, such as supernovae. Lovell had been studying high-energy particle physics as part of his research in Manchester, when he was drafted overnight into radar science. He knew that these cosmic ray showers existed, and might reflect radio waves to some degree, but he would have to wait to test his idea.

Immediately after the war, in 1945, Lovell repurposed leftover military radar equipment to set up his first radio telescope in a courtyard in Manchester. However, the built-up city was too noisy with commercial radio communications for sensitive radio astronomy observations, so Lovell soon moved operations to a quieter site on disused farmland south of Manchester, later known as Jodrell Bank. To Lovell's astonishment, his equipment didn't just work; it exceeded expectations. Lovell had rigged up his system such that every time a radio signal was detected, his machinery would emit a 'ping', just like the radar equipment used in the war. His equipment pinged and pinged and pinged, far more than could be explained by cosmic rays alone. His friend Stanley Hey suggested that this was no cosmic ray shower, this was a meteor shower.

Knowing nothing much about astronomy, Lovell sought expertise elsewhere but quickly found out that most astronomers considered meteors beneath their notice. And so it was that Lovell convinced an amateur astronomer, Manning Prentice, to join him at Jodrell Bank for the next big meteor shower. Now, the site hosts one of the world's largest steerable radio telescopes, the 75-m Lovell Telescope. At that time, though, it was just a shed in a field, with a small antenna on the roof.

Lovell's leftover radar equipment, once used to track enemy aircraft, now revealed the paths of a different kind of enemy: space rocks flung towards us from the asteroid belt. Meteoroids, as they blaze through the atmosphere, strip the electrons from the atmospheric particles they encounter, *ionising* them and creating trails of charged particles. These trails reflect any incoming radio wave, whether the intentional, investigative efforts of radar, or the constant ambient signals from TV and radio broadcasts. Whatever the incoming signal, it was the reflections that mattered, as they rebounded towards Earth, and Lovell's antenna. It is like trying to scope out the size of a cave by yelling and listening for your echo: it doesn't matter what you shout, or that someone else does the shouting, as long as you listen for the echo.

Detecting the meteor trails of small space rocks reaching the atmosphere was the beginning of a big breakthrough. It would lead to radio astronomers using their radar systems to look further out, seeking echoes from asteroids themselves, long before they reached Earth. Asteroid tracking had previously relied on spotting them reflecting sunlight at night, much like the Moon. The ability to detect these objects using radio astronomy, day or night, and in any weather, would be a major leap forward in defence strategy in the years to come, after Lovell confirmed his findings.

Prentice and Lovell spent their nights correlating the optical and radio observations, to make sure of the explanation. Prentice would lie back in his deckchair with his celestial map, torch and notebook at hand, and yell when and where he saw the streak of a meteor. Lovell, in the nearby shed, would listen on his radio equipment, confirming that each yell from the optical astronomer was accompanied by the sound of a 'ping' in his headset. Again and again, a yell accompanied a ping. What a moment that must have been, hearing such a tangible connection between optical and radio. I walked into the control room of an amateur radio astronomy group in England once, just as one of these meteor pings sounded out. It was broad daylight and so I got a real thrill knowing I had heard a shooting star that was otherwise invisible. The amateur astronomer I was visiting, however, turned the sound off, remarking that 'once you've heard one ping you've heard them all', which rather speaks to the prevalence of meteors.

Lovell was bitten by the radio astronomy bug and became a leading figure, if not *the* leading figure, in the UK. Once he realised the promise radio astronomy held for exploring the Universe, he immediately set about designing and building the 75-m steerable radio dish that bears his name. Over seventy years on, it is the third-largest steerable radio telescope in existence, and it is still used for research. It is a magnificent sight, set in beautifully cultivated grounds. Lovell was a keen botanist, and from the 1970s onwards, he and his team planted over 3,000 trees and shrubs from seeds and saplings originating across the world in the area surrounding the telescope. Lovell was dedicated to his radio astronomy work, it became his life, but the beautiful horticulture shows he was spending a lot of his time looking down too.[2]

The Lovell Telescope was almost not built at all. Lovell's passion, ignited by the ping of meteors, had got him so far in securing funding, but repeated delays and parliamentary investigations into stretched budgets almost tanked the project in the 1950s. In the end, radio astronomy in the UK was saved by an unlikely source: the Russians. When the world learned that the Soviets had launched the first satellite in 1957, a beeping potato of a thing called *Sputnik*, there was widespread fear that this marked the beginning of a new era of space-based weapons and cold war escalation. Lovell realised that his radio antenna at Jodrell Bank was one of very few technologies on Earth capable of tracking these new threats . . . and he did not hesitate to use this information as leverage.

The UK government swiftly dropped their complaints and released the funding needed for Lovell to attach some radar equipment to his telescope. Sure enough, Lovell was able to send out beams of radio waves and hear the echoes from *Sputnik*'s rockets. The military soon caught up with their own networks of tracking radar telescopes, and Lovell was left a hero with a shiny new toy with which to investigate the cosmos.

Radio astronomers have often benefited from the paranoia sparked by international conflicts. The 305-m dish at Arecibo was also built for defence purposes, intended to track airborne nuclear missiles. It is an uncomfortable truth, as we might prefer a more peaceful, scientifically motivated source of support, but radio astronomy's role in national security has undoubtedly advanced the field.

While radar has been an important part of the development of radio telescopes, not many radio telescopes amongst the hundreds worldwide are capable of *planetary* radar, the method of bouncing radio waves off celestial objects. Most pioneers like Lovell took leftover radar equipment, designed for detecting enemy aircraft, and turned it into instruments that probed the cosmos by passively receiving the radio waves emitted by stars, galaxies and cosmic phenomena like pulsars. For decades, these instruments focused mostly on listening, with no need to send out signals.

But now that we are aware of the asteroid threat, and we actually have the technology to do something about a collision, radio astronomy facilities have begun adding radar capabilities. Instead of listening passively, as we do with distant galaxies, we are broadcasting signals, using radar to map and monitor asteroids that might pose a risk. In this way, radio astronomy has returned to its origins: protecting us from incoming dangers, not from the skies of wartime Europe, but from the far reaches of space itself.

Listening for the pings of a meteor is one thing, but if we're talking planetary defence, we need to spot an incoming asteroid long before it begins to travel through the atmosphere. Asteroid detection is a multi-wavelength and multi-national effort. When we emit radio waves we can spread our power across the skies, sending a weak signal over a broad, nearby area, or in a narrow, 'pencil' beam that is strong enough to reach further, albeit over a small area. The former is used rarely, when we don't really know what we're looking for, while the latter is used when we know precisely where our target is, be it asteroid or moon crater. Radio telescopes cannot survey wide areas out to large distances efficiently because shooting narrow beams in the dark is unlikely to result in a discovery.

Optical telescopes, in comparison, are just passively collecting reflected sunlight. Without the need to send continuous probing signals, optical telescopes can efficiently survey the whole sky night after night. Then, it is simply a game of Spot the Difference: the new light in the sky that doesn't line up with any catalogued object is most likely the first hint of a new asteroid. It doesn't even require advanced technology. Amateur astronomers, with their exceptional knowledge of the night sky, can and do spot them. The difficulty comes in making sure we can keep track of them.

While optical telescopes act as scouts and make the initial discovery, radar data is essential for understanding whether that asteroid is heading our way. The Doppler shifting of radio waves as they reflect from the surface of the asteroid gives us a measure of the movement: whether the asteroid is travelling towards or away from us, and how fast. If our echoes are of slightly longer wavelength, it is moving away; if they are shorter wavelength, the asteroid is moving towards us. Optical photographs taken over successive nights can offer a glimpse of how an asteroid is moving across the sky, but only if it's on the night side of the Earth and the skies remain clear. It is a little like watching a distant aeroplane cross the horizon. You can work out it is going left or right across the sky, but if it is really far away you couldn't say whether it was moving towards or flying away from you. By the time you might see it getting bigger or smaller, it will have moved out of view, the clouds might have rolled in, or night may have fallen.

Tracking a newly discovered asteroid requires cooperation and swift communication between optical and radio observatories. As soon as an optical observation is announced, radio telescopes can direct a narrow, powerful beam of radio waves towards that location and gather the vital extra information on direction and speed of movement. This follow-up must be fast, within a 'radar window' of about five days. After that time, the asteroid drifts too much along an uncertain path, making it almost impossible to relocate. Without that quick response, we could all too easily lose track of a civilisation-ending asteroid, especially if it slips into the daylight side of Earth, hidden from our optical view.

When we do find a new asteroid and have enough radar information to accurately chart its orbit, we can assess its threat level. Those that pass close to the Sun are termed near-Earth asteroids (NEAs) or near-Earth objects (NEOs). As of 2025, 'close' is defined as within 1.3 times Earth's distance from the Sun, under which definition there are currently over 38,000 known NEAs.*[3] Of these, well over 2,000 are defined as potentially hazardous asteroids (PHAs) according to strict criteria. They must be above 140 m in diameter (a little longer than a football pitch) and have an orbit that will bring them within twenty times the Moon's distance from Earth. That might seem pretty far away, an easy near miss, but over thousands of years orbits can change slightly, due to the effect of a planet's gravity, or a collision with another asteroid. PHAs are close enough that, with a little nudge, they could end up on a direct collision course.

These categories are a bit subjective, especially when it comes to the size restriction.[4] Damage scales with the energy of the event, which grows with mass and with the square of speed, so both size and speed matter, but speed counts far more. Throw a football at someone and it will simply bounce off, whereas a gun can impart huge speed to a tiny bullet to devastating effect. It's the same for asteroids: the size of the rock is important, but so is the speed at which it might hit us.

The Chelyabinsk airburst event in west-central Russia on 15 February 2013 proved an unexpected test of the definition of a

* This number will quickly be outdated, following the activation of the Vera C. Rubin Observatory in July 2025. Within the first ten hours of observation alone, Vera Rubin observed over 2,000 previously undetected new asteroids, including seven NEAs.

PHA.[5] As the city of 1.2 million people got on with their day, a previously undetected 20-m asteroid entered the Earth's atmosphere at a speed of about 40,000 mph. The asteroid heated up quickly and exploded above Earth at an altitude about twice that of a commercial airliner, releasing energy equivalent to more than twenty Hiroshima bombs. The flash briefly outshone the Sun, and the blast shattered the asteroid so completely that most of the meteorites later recovered were only a few centimetres in diameter.* There are no records of anyone getting hurt by a direct strike, but there were still plenty of injuries and damage to property as a result.[6] The bright flash was painful for those unfortunates looking directly at it. Those who had not seen the explosion, nor been temporarily blinded, ran to their windows to see what had happened. Minutes later, when the shockwave arrived, buildings shook and the windowpanes shattered inwards. I'd call that hazardous.

The asteroid collided with our atmosphere in the daytime sky, so optical telescopes had no chance of picking it out in the glare of the Sun. Not even NASA knew it was coming. This happens. We have great technology, but it is not perfect, and we are not able to run it such that it is scanning the sky continuously. This was not the first meteorite strike of this size, but it was the first one in the digital age. Footage of the asteroid exploding was captured on car dashcams and phones and quickly shared around the world. The interesting part of this story is that the asteroid was far from the size needed to flag it as 'potentially hazardous'. The asteroid was only 20 m in diameter, far below the 140-m threshold and so, even if we had seen it, we may not have flagged it for follow-up. Clearly, these categories need reassessing.

It is in everyone's interest to detect an incoming asteroid when

* I own one of the fragments, being the kind of person that joyfully parts ways with £150 just to own a piece of space gravel. I came across it in a fossil shop, which is where you often find meteorites. Odd, really, when one considers the awkward history between dinosaurs and asteroids.

it is still very far away, because any intervention will take time. Assuming we have realised we have a problem, how does one then go about stopping an asteroid? The method that instantly comes to mind is sending up a team of oil miners in a rocket to drill down into the centre of the asteroid and drop a nuclear bomb inside, cracking it into two pieces that will split off and miss Earth, all to a background soundtrack of Aerosmith. This is the plot of *Armageddon*, a film from the early 2000s which I watched many times as a teenager. I know one planetary defence astronomer who admits to watching it annually even now, just to keep up his motivation.[7] When you think about it though, doing that eighteen days, *days*, before impact does seem a little far-fetched. To split the *Armageddon* asteroid in two, an asteroid 'the size of Texas' (1,200 km), and deflect it enough with less than a week remaining, would have required incredible firepower. Even if such a nuclear bomb could be manufactured, what you have then done is turn one massive asteroid into two smaller radioactive asteroids, each several times larger than the asteroid that killed the dinosaurs. The best-case scenario is that it would rain radioactive rocks from the sky.

But with more notice than eighteen days, can we really do anything about a collision? The answer is yes, and in a more peaceful manner too. One method is to manufacture a gigantic 'solar sail' made of a reflective foil. Then, after we have sent a space probe to attach it to the asteroid, the stream of photons from the Sun would provide a radiation pressure that gently fills the sail, slowing the asteroid down and changing its course.[8] Another method is tethering a smaller asteroid to the larger, dangerous one, changing its centre of mass and causing it to move to a different orbit.[9] Any method, whether nuking, sailing or tethering, would require tens of years, if not *centuries* of advance warning. We do have options, though, and they really could work if we spot the threat early enough.

As well as telling us if an asteroid is on a collision path with Earth, radar observations can help us plan this pre-emptive strike. The craggy, boulder-strewn surface of an asteroid presents a challenging site on which to land a space probe, in order to attach a solar sail or a tether. As we have seen with planetary surfaces, radio light is reflected in different directions and polarisations according to the asteroid's terrain, such that we can map it like the surfaces of our nearby planets. The Doppler shifting of the radar signal can also inform us about the rotation period of the asteroid, just as it can with planets, an important factor in deciding how to approach the asteroid safely. Remarkably, using ground-based radar facilities, we can create high-resolution 3D models that are as accurate as optical images captured by space missions that have taken years to reach the asteroid.[10] Should Armageddon come, we won't have time to wait decades for a good camera to reach our enemy.

The NASA *OSIRIS-REx* mission (Origins, Spectral Interpretation, Resource Identification, and Security – Regolith Explorer) launched in 2016. Radar observations by Goldstone and Arecibo were crucial in mapping the asteroid's movement and producing the pre-arrival shape and spin models that were central to mission planning. *OSIRIS-REx* arrived at Bennu in 2018, collected samples of the surface rocks in 2020 and returned the material to Earth in September 2023, providing valuable insight into the composition of asteroids and proving that we could land on an asteroid successfully.

Scientists have even begun testing the methods we hope will change the orbit of an incoming asteroid. The *DART* mission (Double Asteroid Redirection Test) was a NASA planetary defence mission launched in 2021. *DART* successfully impacted a small asteroid called Dimorphos, which is orbiting a much larger asteroid called Didymos, in September 2022, altering its orbit. While radar played a supporting role in mapping the asteroid-moon, it was less crucial this time since this was a kamikaze mission for the *DART* probe. Observations taken before and after the impact revealed that Dimorphos' orbital period around Didymos shortened by around thirty-three minutes. It marked the first real-world test of our ability to deflect an asteroid, and it worked. But it was a small-scale rehearsal. No matter how powerful our punch, it's useless if we throw it too late.

Our foray into the outer Solar System has taken us from floating past the seemingly peaceful rings of Saturn, amongst the icy entrails of a long-dead moon, to weaving through the asteroid belt and ducking for cover when rocks are forced out towards Earth. The speeds at which these asteroids hurtle through space and the sheer number of them scattered across the inner Solar System mean that Earth is gliding through a minefield of existential threats, unable to brake or change course. We are reminded annually of the debris that exists out there, as we watch meteors streak across our night sky, and listen to the radar pings of their trails in our headsets, even during daylight.

Defending our planet from a large asteroid strike requires a global effort. Optical astronomy will spot the first glimmer of an incoming asteroid, while radio astronomy can chart its trajectory, assess its threat, and map its surface. Only then can a space-based mission intervene to alter its course or neutralise the danger. Asteroid impacts are a threat that no single nation can face alone; they demand a coordinated, international defence across borders, centuries and wavelengths.

What greater cause for unification is there? Our planet is a blue jewel in a dynamic and fascinating neighbourhood. With just an antenna, we have already seen wonders that we might never have expected: ice on Mercury, caves on the Moon, volcanos on Venus and underground seas on Mars, all encompassed in the violent but nurturing atmosphere of our central star, the Sun. I feel extraordinarily lucky to be one of the tiny lifeforms inhabiting Earth, part of a species that can not only defend itself against space-borne dangers, but harness light to explore the beauty of our Solar System, covering distances we could never hope to travel ourselves.

In part 1 of this book, we have conversed with the gods – Mercury, Venus, Mars, Jupiter, Saturn. But after we have passed two more – Uranus and Neptune – then beyond the Kuiper belt and through the icy haze of the Oort cloud, the gods fall silent. Breaking out of the solar bubble that protects us, our gaze must adjust to a deeper, darker sky. Now we glide past constellations etched across the heavens, within which we will find stories of stars, galaxies and cosmic phenomena more ancient than any Roman god. Far from the Sun, it is cold, dark and lonely, but with our trusty radio antenna lighting the way, it is time to step into the wilderness of the Milky Way.

OUR GALAXY

7

THE MILKY WAY

Why should I feel lonely? is not our planet in the Milky Way?

Henry David Thoreau, *Walden*

Owing to the light pollution in my home country, I was already twenty years old the first time I saw the Milky Way. I was in Yellowstone National Park in the USA, sitting by a campfire, my toothpaste safely locked away in a bear-proof cabinet. There was conversation going on around the fire, cheerful laughter and the hiss of beer cans being opened by my fellow campers. Even with flickering firelight right beside me, the sheer number of stars visible in that gorgeous ribbon overhead took my breath away. I was two years into a theoretical physics degree and suddenly my pen-and-paper astronomy courses didn't seem so theoretical any more. I could see the galaxy, our galaxy! To my eyes, the stars were so densely packed they blurred into one another, making it hard to single out any one star. That night around the campfire, space ceased to be abstract, and the Milky Way became home.

Even in that remote mountaintop location, I could see at most a few thousand stars, a sliver of the hundreds of billions that make up the Milky Way, limited as we are by the human eye. That's like downloading a high-resolution image of the *Mona Lisa* and finding you can only see 10 out of the 64 million pixels. Most of us will rarely glimpse even this small sample of our galaxy due to the light pollution from homes, offices and streetlamps. Light pollution from more natural sources like the Moon is a significant problem too. In the day, the Sun dominates completely, concealing all other stars, but even at night its influence persists. The Moon is 400,000 times fainter than the Sun, yet that is enough to wash away thousands of faint stars, stealing the full splendour of the Milky Way from view.

No lover of art would be satisfied with ten pixels of Mona Lisa's enigmatic smile, and neither is the astronomer content with the few pinpricks of light our eyes can gather from the vast canvas of the Milky Way. And so, we must find ways to escape the limits of our bright environment and our crude eyes.

In 1977, NASA launched *Voyager 1* and *Voyager 2*: two small probes armed with long antennas and big dreams. The probes were designed to take advantage of a rare planetary alignment that occurs only once every 176 years, when Jupiter, Saturn, Uranus and Neptune all draw close enough together to allow a single mission to visit them one after the other. The *Voyager*s were able to gain speed and change direction by passing close to each planet, using gravity assists to 'catapult' them onwards to the next planet on their itinerary, first around Jupiter to reach Saturn and then, in *Voyager 2*'s case, Saturn to Uranus and Uranus to Neptune. One by one, the outer gas giants and ice giants filled the frame of the onboard camera: a storm whirling across Jupiter, the faint rings of Uranus, clouds streaking across Neptune. Even the tiny moons got a close-up: the cratered surface of Rhea could be our own Moon, it is so detailed, and the cracked surface of Miranda looks like the rift valleys of Iceland, tortured and jumbled. But as quickly as they had loomed into view, each planet slowly drifted away. As the *Voyagers* flew beyond Neptune, they entered the Kuiper belt, a ring of debris much like the asteroid belt. One might imagine images teeming with the hundreds of millions of icy comets there, but in truth, these comets are spaced so far apart that even the gap between Earth and Mars would seem cosy in comparison. The *Voyagers* captured only darkness: a portrait of empty space.

Theirs was a one-way ticket. No physical samples were taken; everything, data and images alike, was sent back via radio waves, kilobyte by kilobyte. The trajectories of the probes at the end of their planetary tour meant that, after twelve years, instead of coming home, they headed outwards, into the wider Milky Way.

Yet the Sun's influence stretches so far that it would take decades more to pass beyond its territory and into true interstellar space. That boundary is defined as the point where the outward flow of the solar wind is overcome by the incoming cosmic rays. Both are streams of charged particles, but while the solar wind pours out from the Sun, cosmic rays originate from the depths of the Milky Way. It took two years for the *Voyagers* to get to Jupiter, three years to get to Saturn, nine to get to Uranus and twelve to get to Neptune, and still the solar wind was going strong. Only after thirty-five years did the instruments detect a severe drop in the number of charged particles from the solar wind and a sharp increase in galactic cosmic rays. In 2012, *Voyager 1* became the first human-made object to cross the boundary into interstellar space.[1]

We are a planet sitting in the outer atmosphere of our star, the 'heliosphere'. This bubble of particles carried by the solar wind acts as a giant force field, pushing back on incoming cosmic rays which, if they reached Earth in great numbers, could be a potentially fatal source of radiation. Leave the heliosphere, and you enter interstellar space. But you also leave the shelter of the Sun.

By the time the *Voyager* probes had reached the interstellar frontier, 18 billion km from Earth, their radio communications took sixteen hours to reach us. For context, light from the Sun takes 'only' eight minutes to reach Earth. Once outside the heliosphere, the Sun has shrunk in apparent size to become just another star – though still brighter than the rest. There are no planets to look at, and no asteroids to avoid. Interstellar space contains just a thin scattering of hydrogen and helium atoms, and a sprinkling of heavier elements such as oxygen, iron and gold expelled from supernovae. While still not a perfect vacuum, interstellar space is far emptier than anywhere within the Solar System. Earth's atmosphere contains around 10 billion trillion molecules per cubic centimetre. A similar volume of interstellar space would contain just 1 atom.

Despite travelling at blistering speeds of over 60,000 km/h, the void between the stars is so immense that the probes must cruise through this featureless expanse for millennia. *Voyager 1* will come within 1.7 light-years of star AC +79 3888, located in the constellation Ursa Minor, in 40,272 AD. A few hundred years after that, *Voyager 2* will pass similarly close to star Ross 248, in the constellation of Andromeda. These are not the closest stars to the Sun (Proxima Centauri is half the distance) but when one is speaking in units of ten thousand years, it hardly matters. Long, long before then, the *Voyager* probes will have run out of power and will be unable to send us data. Even if they had the power, they are so far away that their radio signals would be too weak for us to detect. Space, as Douglas Adams sums up in *The Hitchhiker's Guide to the Galaxy*, is 'big. Really big.'

The *Voyager* probes have serenely glided through our Solar

System traversing a timeframe that spans presidential careers, from Jimmy Carter to Barack Obama and beyond, and musical tastes from Fleetwood Mac to Drake. While our society has evolved both politically and culturally (or devolved, depending on your point of view), the *Voyager* probes are frozen in time. They are envoys from humanity as it existed in 1977, constrained by the engineering, scientific and computing power of that time. As their power sources diminish, NASA has had to begin closing systems down one by one. The cameras were turned off in 1990 to save both power and the onboard memory, all 70 kilobytes of it. Remarkably, the *Apollo* missions used a similarly modest amount of memory to run the systems that navigated them to the Moon. In comparison, a modern-day mobile phone has over 3 million times that amount.

That's one giant leap for technology, but no amount of improvement on 1970s technology is going to help us bridge interstellar space. Even if we launched a smartphone at the speed of light tomorrow, it wouldn't reach Proxima Centauri for over four years. And that's just the closest star. What if we want a wider perspective of our galaxy, not a picture taken from within, but a galactic selfie from the edge? Even at the speed of light, it would take our smartphone *25,000 years* to reach a tourist vista in the closest edges of the Milky Way. And this is an unrealistic, best-case scenario. While we may one day have the technology to create a battery that can last that long, light-speed travel belongs solidly within science-fiction.

We have dozens of spacecraft other than the *Voyager*s, all currently exploring the Solar System, providing high-resolution images of everything from Mercury to Saturn's moons and even the surface of Pluto. The Hubble Space Telescope and the JWST don't travel, but they explore by gathering the faint light of galaxies billions of light-years away, sending us images with a level of detail the human eye is not equipped to appreciate. In contrast, every single image you have seen of the Milky Way – those face-on views with majestic spiral arms – are artist's impressions only, based solely on what we have observed from within.

If our galactic self-portrait was informed only by optical observations, it would cover a small area indeed. Radio waves, in contrast to waves of visible light from stars, are emitted from the gas underpinning the entire structure of the galaxy, and can travel from far-away regions, cutting through the interstellar dust with ease. And so, while radio astronomy may not provide an ultra-detailed star-by-star map like nearby optical observations, it can give us the broad brushstrokes needed to understand what kind of galaxy we live in.

A galaxy is a collection of stars, gas and dust, all held together by gravity. There is another very important, yet elusive, component called dark matter, but that's a mystery we will unravel in a later chapter. Observations by modern telescopes have revealed a remarkable diversity in the shape, size and colour of galaxies. Broadly, we separate this galaxy zoo into three categories: elliptical, spiral and irregular. Elliptical galaxies are spheroidal, made up mostly of older stars, with a central concentration that gently fades in every direction. Spiral galaxies, in contrast, have a bright, central bulge similar to the form of an elliptical, but surrounded by a flattened disc of gas and stars rotating in an orderly fashion. Within these rotating discs, gas is drawn together into spiral arms, forming high-density regions that are fertile grounds for star formation. Irregular galaxies lack the tidy structure seen in spiral or elliptical galaxies, often shaped by low mass, slow rotation or violent interactions with nearby galaxies. In fact, many elliptical galaxies are thought to be the aftermath of similar collisions, where orderly, spiral galaxies are stirred together into a more rounded form.

The Milky Way is a spiral galaxy: hundreds of billions of stars, all gathered in a grand celestial structure with a glowing central bulge and a swirling set of spiral arms. Earth is located 27,000 light-years from the galactic centre on the Orion spur, a minor spiral arm nestled between the two major spiral arms, Sagittarius and Perseus. Our view towards the centre is of the other spiral arms superimposed, creating a tangled ribbon of light. Galileo Galilei was the first to pick out individual stars in the misty Milky Way, taking advantage of a recent invention: the telescope. He sketched the locations of the stars, invisible to the naked eye, and published the first maps of the Milky Way in 1610, in a beautifully named paper called 'The Starry Messenger'.[2]

While the invention of the telescope provided a huge leap forward for astronomy, it was a difficult technology to master and improve, and so those first star charts were not much more than sparse dot-to-dots. For example, whereas modern amateur telescopes can resolve over 1,000 member stars of the 'Beehive' star cluster, Galileo could only see 40. Clear observations required the grinding down of glass lenses to extraordinary smoothness, and it would be quite some time until such industrial techniques were invented.

Galileo did have one distinct advantage though: dark skies. Over hundreds of years, human scientific endeavour has slowly but steadily increased our understanding of optics and engineering such that bigger, more accurate, and cheaper telescopes could be manufactured. But alongside this progress came other breakthroughs, namely electric lights.

Since 2001, astronomers have rated the darkness of the night sky on the 'Bortle Scale' after the astronomer who created it, John Bortle. A location with a Class 1 rating on the Bortle scale is as dark as they come, usually mountains and deserts so remote that starlight casts shadows. On the other end of the scale, Class 9 locations are inner-city skies where you can see the Moon and not much else. Galileo took his observations from Padua, in northern Italy, a city of 20,000 inhabitants in those times. Lit only with oil lamps, Padua in 1610 would have had a Bortle rating of Class 2, the same light environment I experienced at 2,000 m above sea level in Yellowstone National Park. Today, the city would fall between a Class 7 and Class 8, and a modern-day Galileo would struggle to see more than a handful of stars. In Padua, as in most towns and cities, human progress has rendered the Milky Way entirely invisible.

Astronomers moved their observatories further and further from civilisation, first across the globe and then, following the great leap of the space race, away from that globe entirely. The *Gaia* spacecraft, launched by ESA in 2013, mapped the positions of around 2 billion of our closest stars, from a position 1.5 million km away from Earth. *Gaia*'s maps are exquisitely detailed and, for now, represent the cutting edge of stellar cartography. Yet they chart only 1 per cent of the galaxy's stars. The rest remain hidden, drowned out by brighter stars or veiled by dust. In mapping the Milky Way with optical data, we are attempting a complex dot-to-dot, but with most of the dots printed in invisible ink. With only the dots we can see with our eyes, we could trace the outlines of galaxies of all sorts of shapes, with any number of spiral arms. To work out the true shape, we need to get to the dots beneath the dust.

Astronomical dust is like the household dust we are all familiar with, in that it gets everywhere and we'd often rather it didn't. Stars create their light by fusing hydrogen atoms into progressively heavier elements, until that hydrogen fuel runs out. At this point, many of the stars puff outwards, or even explode in supernova, expelling those fine, solid particulates of heavier elements into their surroundings. This dust mixes with the gas to once again collapse into a new star, perhaps some planets this time, which collide and create more dust, before this newer star explodes, adding more dust and so on. With each cycle of creation and destruction, tiny solid particles fill space like the haze one sees during a fireworks show, absorbing the light, dimming and hiding the starbursts in the sky.

This mask of light-obscuring particles absorbs a lot of starlight, especially in the bulge and disc of the galaxy, where star formation is most rife. For optical astronomers, especially those who want to study our own galaxy, it is a real problem. When we see the Milky Way from Earth, these dark channels of dust are not subtle: they are arguably just as striking as the 'milky' haze of stars. Indeed, in many First Nations Australian traditions, it is the dust that defines the galaxy, forming the shape of a bird they know as the 'Emu in the Sky'. I know these cloudy areas by a different name: 'The Zone of Avoidance'. If we were to still draw our maps on parchment, it would say 'terra incognita: astronomers keep out'. We can look above and below this zone of the Milky Way and observe stars in galaxies billions of light-years away but, within that zone, we can fail to see stars in the spiral arm next to us. It all seems rather hopeless, until you consider that dust is a poor match for longer wavelengths of light.

There are two ways to approach the problem of dust at home. One is to wage a constant war by spending hours sweeping and polishing, much like the efforts of optical astronomers who painstakingly clean their data of the effects of dust in the hope of seeing starlight. The second way is to look at dust with different eyes and learn to not care that it's there. One look around my house will convince you that it is this second way that I favour, both at home and as a radio astronomer. Radio astronomers avoid straining to see stars through the dust, and instead observe the scaffolding of gas on which those stars hang. Not only is the gas far more spread about than the stars (think broad washes of colour vs that dot-to-dot), but the radio waves bypass the inconvenience of dust, careering through unhindered.

Radio waves trace the cold hydrogen gas spread throughout a galaxy: vast reservoirs of star-forming material that produces neither optical nor infrared light. Along the spiral arms, density waves act like slow-moving traffic jams, causing the gas to clump together. These crowded regions spark bursts of star formation, yet even after the stars form from some of the gas, there is plenty left over. When we look up at the night sky, our eyes see a galaxy made of pin pricks of light – a dot-to-dot masterpiece in which we have drawn our own lines, creating the constellations. In radio though, the galaxy is more akin to van Gogh's *Starry Night*, where the stars sit within swirls of colourful gas, covering the entire night sky.

This gas is the building material for all the stars we see. It is that first wash of paint on a canvas, upon which all other colours are built up over time. It is everywhere throughout the Milky Way, but while dust shrouds the structure of the galaxy, the gas reveals it. While optical observatories like *Gaia* have produced an intricate portrait of our local neighbourhood using the stars alone, radio observations have filled out the surrounding landscape, painting a vibrant scene so complex that one can see something new in the background every time. And it is a background that isn't limited by perspective, that has no vanishing point. Because the radio light the gas emits travels through dust with ease, this galactic structure can be mapped in detail even out to the *other side* of the Milky Way's bulge, a region so dense with stars and dust that it is impossible for any optical telescope to look beyond it.

To map the gas across the galaxy, we use the radio signals emitted by hydrogen, the simplest and most common atom there is. A hydrogen atom is made up of a proton, which has a positive charge, and an electron, with a negative charge. The electron in a hydrogen atom can exist in different 'energy states' where normally it can be found in the 'ground state', the lowest energy level, the equivalent of sitting around and watching TV. However, our lazy electron can move up to one of the higher 'excited states' by bumping into other atoms and receiving a 'boost' from their energy, like being given a cup of coffee. Maintaining that higher-energy state is difficult, and electrons naturally prefer to be in the ground state. They will expend their excess energy by emitting photons as they return to the sofa of the ground state, traversing down the energy states in steps, leaps and jumps.

In hydrogen, a tiny quantum tiptoe between energy states produces photons with a wavelength of 21 cm – firmly in the radio part of the spectrum. Hydrogen makes up about 90 per cent of all atoms in the Universe (not counting dark matter). It is *everywhere*, so 21-cm radiation is one of the most important diagnostic tools in astronomy, lighting up the darkest corners of our galaxy, where optical tools are useless.

This emission length, 21 cm, is so fundamental to science that it is referenced on the gold-plated, copper disc stored onboard each of the *Voyager* probes.[3] The records are encoded with music, examples of human spoken language and images from Earth, while the aluminium cover is etched with instructions for constructing a record player and a map of our position in the Milky Way. There is also a diagram of a hydrogen atom emitting a photon, to set a unit of length scale. It is all very well stating that a human is about 180 cm tall, but what is a centimetre to an alien? The wavelength of light emitted by hydrogen atoms will be the same in any star system, so any alien with a basic knowledge of atomic physics can use it to compare our height with its own. While mobile phone and television signals cover a broad range of the radio spectrum, the 21-cm band is protected entirely for astronomers, by writ of the International Telecommunication Union, to prevent a radio show from drowning out the whispers from the Milky Way. When an astronomer tunes their telescope to pick up 21-cm wavelength light, they are tuning in only to the radio galaxy.

The first radio map of our galaxy was produced in 1944, by a disgruntled amateur astronomer. Grote Reber was a newly qualified electrical engineer in Illinois when Karl Jansky, also an electrical engineer, published his work on the 'cosmic hiss' coming from Sagittarius. Reber was captivated, but his work in a radio equipment company didn't exactly allow him to follow up the research himself, especially since, at the time, the concept of a 'radio astronomer' was non-existent. Reber made enthusiastic enquiries to academic institutions and was surprised when they showed little interest in advancing Jansky's research. Reber even applied to Bell Labs, where Jansky still worked, but he was rejected, as the hiring freezes following the Great Depression took hold.

Disillusioned with academia, but undeterred from his own curiosity, Reber took matters into his own hands: 'I consulted with myself and decided to build a dish'. Reber took the summer of 1936 off work and, at twenty-six years old, spent one-third of his annual salary constructing the first purpose-built radio telescope in the world. He didn't even do the obvious thing and replicate Jansky's wire and wood 'merry-go-round'. Reber's design, a dish in the shape of a parabola, was the first of its kind, and became a model for future radio telescopes thanks to its ability to gather and focus light onto a central antenna more efficiently.

It was an impressive structure: a tiltable 9.6-m-diameter dish atop a revolving platform that utilised wheels from a crane and the rear axle of a Ford Model T. There were no state-of-the art materials or industrial techniques, just trips down to the hardware store for sheets of galvanised metal, wooden rafters, and over 1,600 screws. The location was makeshift too: Reber built the telescope in his mother's back garden, in the town of Wheaton, Illinois.[4]

The locals were bewildered. The neighbours feared it was a device able to change the weather. Local pilots rerouted their planes to avoid it and school kids routinely used it as a climbing frame when he wasn't looking. On Mondays, it was used to hold up his mother's washing line.

Once the telescope was constructed, Reber returned to his day job, reserving his astronomical pursuits for night, not because he needed dark skies, but because he needed quiet skies. During the day, his suburban environment was overwhelmed with signals from home radio sets and even interference from car ignitions. It took a long time to work out the kinks, but eventually, in 1939, Reber successfully replicated Jansky's observations, detecting the same cosmic hiss.

Reber's breakthrough was initially met with scepticism, with some scientists even dismissing it as a hoax. Reber ignored the slight and began steadily scanning the whole sky, creating the first-ever maps of the radio sky. Just as terrain maps show Earth's elevation with contour lines, where closer lines mean a steeper ascent, Reber's maps used lines to represent varying intensities of radio waves. The tighter the lines, the brighter the radio emission in that region. While Jansky's discovery had served as a signpost, pointing towards a mysterious source of cosmic noise, Reber's map was like an Ordnance Survey chart, showing how this 'cosmic hiss' varied across the whole sky.

Unlike Jansky's rudimentary equipment, which could only pick up the general direction of close-by cosmic signals, Reber's set-up was sensitive enough to detect radio waves from far greater distances and pinpoint them with some accuracy. A dense knot of contour lines within the Sagittarius constellation marked the borders of the galactic centre, which would later be recognised as coming from Sagittarius A★, the supermassive black hole at the centre of the Milky Way.* But Reber also revealed two additional

* The letter 'A' marks the bright diffuse radio emissions, while the added asterisk specifies the compact area near the black hole.

areas of high intensity: Cygnus A and Cassiopeia A, the first extra-galactic radio sources ever detected. Today, we know Cygnus A as a powerful radio galaxy and Cassiopeia A as a supernova remnant, but at the time, they were intriguing points of interest on his map, islands of bright radio emission no one could yet explain.

Upon Reber submitting these groundbreaking results to the *Astrophysical Journal*, the editor, intrigued but wary of a submission with no academic affiliation, made the highly unusual move of sending a delegation of astronomers from the prestigious Yerkes Observatory at the University of Illinois to examine Reber's equipment. They could find no fault, and his paper was approved.[5] Even so, it still took some time for the astronomical community to reconcile the idea that an outsider, an *amateur*, had shown more curiosity and tenacity then they had.

Jansky had remained an electrical engineer who took note of the skies as part of his job to excise the noise from telephone calls, but Reber made the study of the Universe in radio waves firstly a hobby, and then his life's pursuit. Alone in his mother's garden, Grote Reber was the world's first radio astronomer, and indeed the world's *only* radio astronomer until after the Second World War.

As Jansky and Reber's results became more well-known and well-respected, there was a rush to observe the radio sky. Those with a physics background could engineer the equipment but they didn't have a clue what they were pointing at. Those with an astronomy background knew what they were looking for but couldn't create their own radio equipment. In these first years, academics could offer only up to one half of the skills of a true radio astronomer: they could understand the experiment, or they could understand the results. One man in the world could do both. Reber suddenly found himself in receipt of several academic positions, but he declined them all, having never got over the treatment he had been afforded. He remained an independent 'amateur' astronomer his entire life.

In 1944, Reber had scanned the Milky Way and mapped how the radio light varied across the visible ribbon. The brighter the 21-cm signal, the more gas we can infer exists in that direction. But, as novel as his telescope was, it could not give much detail with regards to the structure within the disc. As we saw when employing radar to explore the planetary surfaces, radio data also contains Doppler information on how fast an object is moving, whether that object is a planet or an atom. If a cloud of hydrogen atoms is moving towards us, then the 21-cm light wavefronts will pile up and we will observe the light at a higher frequency, or equivalently at a wavelength very slightly shorter than 21 cm.

In 1952, astronomers used the Doppler characteristic of radio data to deliver the first solid proof that we live in a spiral galaxy. The shape of our galaxy seems self-evident to us now, but that is only because we have grown up with the answer. From our vantage point in the galactic disc, we view the Milky Way edge-on. The stars in all the spiral arms blur together, smudging out any obvious pattern. Yes, the ribbon of light across a dark night sky indicates a disc, but that doesn't necessarily denote a spiral galaxy. There are irregular galaxies with discs and no spiral arms to speak of, for example. In the early 1950s, we still weren't sure exactly what kind of galaxy we were calling home. A radio telescope based in Sydney, Australia, gave us the answer.[6]

It was a great time to be interested in space. A popular fascination with space and its exploration was beginning to form, with sci-fi authors such as Ray Bradbury and Arthur C. Clarke exploring the idea of alien life. Films about aliens abounded, and Buck Rogers and Flash Gordon had made their debuts in comic form. While the space race had not begun in earnest – that would happen in 1957 with the launch of *Sputnik* – it had not escaped the imaginations of military commanders and little boys alike that the missiles developed during the war might be used as a way to propel man to space.

At that time, almost all astronomy was optical astronomy. We knew Venus had a thick cloud cover, but without radar we didn't know what lay beneath. We knew the Earth sat in a disc of stars, far from the bright centre, but we didn't know for sure there were any spiral arms. We knew there were other galaxies, and the Big Bang theory had recently become mainstream, but the term 'black hole' didn't exist yet. The popular fascination with space was focused on a tiny portion of what was out there to explore: space travel to the Moon, and aliens on Mars. Radio astronomers, meanwhile, had bigger ambitions. They wanted to explore the whole galaxy.

When they had built the means to do so, they saw multiple 21-cm emission lines, Doppler shifted by slightly different amounts.[7] A smooth disc of gas would produce one line, albeit Doppler shifted because the gas will be moving relative to us. But the presence of two or more lines suggests that there are multiple gas structures along that direction, each moving at slightly different velocities. This is exactly what we would expect, since the spiral arms move at slightly different speeds relative to one another. While in both the radio and the optical we see a bright ribbon of stars from different spiral arms superimposed, in the radio we can separate out the contributions and see the underlying structure.

The two Doppler-shifted lines indicated two structures separated by about 10,000 light-years. This gargantuan distance suggested that the structures were indeed two spiral arms of a galaxy. This discovery cannot be overstated. With radio data taken by a simple antenna, we transformed our maps of the Milky Way from 2D portraits to 3D panoramas: we had peered through the dust and detected spiral arms unimaginable distances away.

Over the decades, astronomers have developed technology and software that can pick up ever fainter 21-cm signals with ever more precision. In 2004, astronomers were combing through new data from the Southern Galactic Plane Survey (SGPS). Using radio telescopes in Australia, the survey sought to map out the distribution and velocity of gas in just the same way as their predecessors fifty years before. The SGPS data revealed clouds of hydrogen in particular areas of the galaxy with large Doppler shifts indicating unexpectedly high velocities. Comparing to various computer models, the team determined that the data was best fit by adding an additional spiral arm wrapped around the Milky Way. The Five-hundred-meter Aperture Spherical Telescope (FAST) in China, the largest radio dish in the world, then claimed the discovery of another new spiral arm in 2021, following detection of a structure in a hydrogen cloud 72,000 light-years from the centre of our galaxy.[8] Even now, our galactic home contains mysteries, secret structures invisible in optical light but betraying themselves as a faint glow in the radio. And each time new data flows in, we have to update those artist's impressions of the Milky Way, just as we have been doing since Galileo first sketched the stars he could only see through his telescope.

Every method has its limitations and, while we can map huge portions of the galaxy using 21-cm radiation, the furthest regions of the galaxy are proving too far off for our current technology to detect. Our beautiful, realistic landscape picture dissolves into that underlying dot-to-dot at its edges. But 21-cm radiation is not the only radio wavelength out there, it is just the most ubiquitous, and it was the obvious place to start when radio astronomy was nascent in the 1950s.

In the last few decades, some radio telescopes have moved away from detecting the diffuse ribbons of hydrogen and instead targeted regions of bright emission called *masers*, the microwave version of the more familiar optical *laser*. Lasers are better known because we encounter them in everyday life when, for example, scanning barcodes or watching the latest spy thriller or science-fiction film. Both rely on the same process: *stimulated emission*. If an atom has been boosted into an excited state, a passing photon can trigger it to drop down, releasing a second photon identical to the first. The result is two photons moving on with the same wavelength, in the same direction. A perfectly amplified signal and a very bright one when the process is repeated over and over again. It was easier to demonstrate this at microwave wavelengths, so the first working maser was built in 1953, years before the first visible-light laser in 1960.

Nature makes both too, though space masers are far more common than space lasers. Water masers occur thanks to the excitation of water molecules in the turbulent gas around young stars, and produce light so bright they can be detected from the other side of the galaxy using radio telescopes.

In order to work out how far away these masers are, astronomers use the classic parallax method. Once a maser is located, they take measurements of its position six months apart, so that the Earth is at opposite points in its orbit. By taking observations from two different positions, we mimic the way our two eyes judge depth. You can recreate this effect by holding a finger in front of your face and viewing it first with one eye, and then the other. The closer your finger, the more your finger appears to move relative to the background. Astronomers use the same principle with masers and background galaxies: the smaller the shift of a maser's apparent position, the further away it must be.

The radio telescopes in use for this job can pick out masers from the blur of radio light with astonishing precision: if our eyes could achieve the same resolution, we would be able to look up at the Moon and call out the denominations of two coins placed side by side on its surface. By painstakingly taking these parallax measurements for hundreds of water masers, scientists have constructed a far-reaching model for the spiral arm structure: placing points on the dot-to-dot of our galaxy well beyond what *Gaia* can manage.[9] One survey has detected a maser 66,000 light-years away, all the way on the opposite side of the galaxy.[10] It is this data, alongside the broader brushstrokes of 21-cm data, that forms the basis for every artist's impression of the Milky Way.

Radio astronomers have been refining the cartography of the galaxy steadily over the decades, enhancing their maps at every possible wavelength. Each introduction of state-of-the-art technology and facilities has delivered ever more detailed images, a process reminiscent of the experience at the opticians, where the world gradually sharpens into focus as lenses are swapped in that absurd headset. In 2018, an astounding image was captured by a telescope called MeerKAT, revealing not only the hidden present, but traces of the distant past.[11]

MeerKAT began as the Karoo Array Telescope (KAT), an array of just seven 13.5-m dishes in South Africa's Karoo desert. Thanks to additional funding, it expanded to sixty-four dishes, earning the name 'MeerKAT' (*meer* meaning 'more' in Afrikaans). The name is charmingly apt. Just as the native meerkats bolt upright to fervently survey their desert surroundings, MeerKAT's dishes stand high on the plains, studying the skies intently. But what MeerKAT revealed is extraordinary indeed.

Hubble and JWST reveal the immediate structure of galaxies: stars, dust clouds and the infrared glow of hot gas in stellar nurseries. Radio astronomy digs deeper, adding a view of the galaxy's underlying dynamics, such as cold gas flows, magnetic fields and interactions within the interstellar medium, which collectively fuel and sustain star formation over time. Like the smoky trail across the sky of a plane long-since flown away, the gas preserves evidence of events from hundreds of millions of years ago. We can produce colour-coded images of the different wavelengths, intensities and polarisations. The colour choice is ours, but the variety and richness of the underlying wavelengths are very real.

The MeerKAT image is a cacophony of colour and form.[12]

Filaments of gas following magnetic field lines, otherwise invisible, light up like lightning bolts across the galactic plane. Scattered across the image are spheroidal explosions of gas, remnants of supernovae, that float like ghostly echoes, the shockwaves of ancient stellar deaths still haunting the fabric of space.

Despite the lines and splashes across the canvas, the eye is drawn to the centre of the image where, dazzlingly bright, lie clouds of burning-hot gas surrounding the heart of our galaxy: Sagittarius A*. The gas here erupts from the centre in two billowing structures that reach hundreds of light-years above the black hole. These outflows are evidence of a phenomenally energetic burst of matter that erupted near the black hole a few million years ago. It could be that the black hole swallowed a particularly large amount of gas and burped, or it could be shockwaves from a massive burst of star formation in the neighbourhood of the black hole. The gas is a snapshot of history, and over time, with more data, we can build up the story behind it.

This image is so exquisite that it feels more like art than science, calling to mind Picasso's fragments, Pollock's splatters and Dalí's surrealist expression of time, uniting the ancient past with the fleeting present. More than anything, this image presents a living galaxy, one at work and in flux. It's a startlingly dynamic view, suggesting we live in a galaxy that breathes, evolves and remakes itself over cosmic epochs.

In the end, it's this search for completeness, for perspective, for an understanding of our place in the galaxy that links MeerKAT to the *Voyager* probes, drifting now far beyond the Solar System's edge. They will one day drift close to distant stars, but they will never return their outside perspectives on the Milky Way. Instead, our search for a portrait requires looking at the stars, gas and gravity that surround us.

Perhaps it's a humbling reminder that for all our explorations, we are still creatures of the Milky Way, bound within its arms.

We will always need some tool, whether it be a radio telescope, a space probe or the determination and imagination of a disgruntled amateur astronomer, to navigate our way through it.

On Valentine's Day 1990, American astronomer Carl Sagan asked NASA to turn *Voyager 1* around, just for a moment, so that it could take one last picture of Earth as it whizzed outwards past Neptune. In the image, Earth is contained in just one pixel, a tiny 'pale blue dot'. To quote Sagan himself: 'Look again at that dot. That's here. That's home. That's us. On it everyone you love, everyone you know, everyone you ever heard of, every human being who ever was, lived out their lives.'[13] It is easy to miss. We are a planet surrounded by unfathomable amounts of space, adrift in a galaxy so large it feels impossible to fully grasp the size of it from the inside. It takes radio waves to bring structure to these unknowns, revealing the hidden architecture of where we live, the true shape of our home: the Milky Way.

8

BLACK HOLES

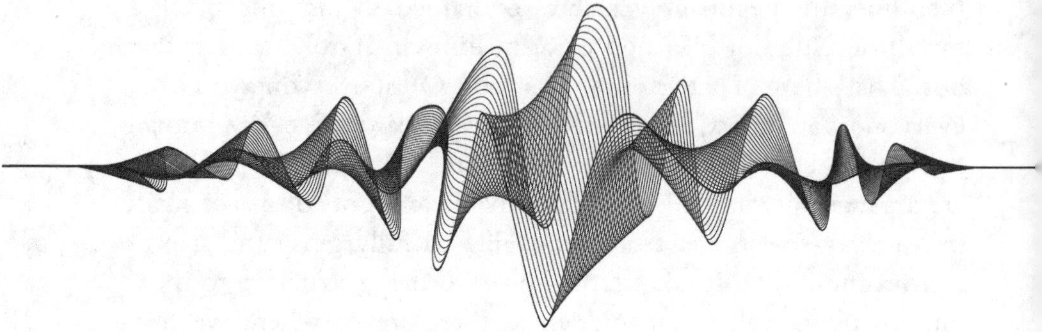

We have seen the gates of hell at the end of space and time.

Heino Falcke, Principal Investigator,
Event Horizon Telescope

The image of the *Voyager* probes sailing through space, threading their way between the stars, is a serene one. These twin emissaries of human curiosity are heading steadily away from the centre of the Solar System, each following its own path. Their missions are ambitious: exploring the outer planets, probing the heliosphere and venturing into interstellar space. Yet I often wonder: if I'd been in charge, would I have been tempted to aim one probe in a more chaotic direction, say, towards the supermassive black hole lurking at our galaxy's core? Is there a more mysterious location in our galaxy?

A black hole is, mathematically speaking, a 'singularity': a point where density, temperature and spacetime curvature all become infinite, following the implosion of a massive star. Surrounding this enigma at the heart of our galaxy is the event horizon, a boundary where gravity's pull becomes so ferocious that not even light can escape. The space within this event horizon is off limits to us experimentally . . . and to some extent intellectually. Our brains cannot imagine an infinite density, or a point in space with no volume. At the centre of a black hole, the dependable rules governing gravity, and those describing the tiniest scales, the quantum world, clash violently, with neither fully explaining this oddest of phenomena.

We cannot hope to explore beyond the event horizon, by any physical means, and we knew this in 1977, when the *Voyager* probes were launched. This is one reason why no one thought to launch one of them towards the centre of the galaxy. But there's also the small detail that it would take a probe hundreds of millions of years to get there; it would arrive long after any human technology (and probably any human society) could hope to be operational. Even if we somehow developed technology allowing us to drastically reduce the probe's journey time, all communication and observation would cease immediately at the event horizon. What would happen inside? Your guess is as good as mine.

All this is like telling a three-year-old they cannot ever open the most enticing present under the Christmas tree. Even to understand the outer reaches of black holes, we've had to resort to metaphorical box-shaking: does it rattle? Does it emit gravitational waves? What clues can we glean from the doomed gas spiralling inwards?

Black holes captivate the imagination like no other celestial, or terrestrial, phenomenon I know of. Every outreach astronomer, whatever their research area, must be prepared to field endless questions from fascinated children regarding black holes. What are they made of? Are they entrances to wormholes? We often cannot provide satisfying answers, and the truth is, astronomers have many of the same questions. Even Einstein struggled with the concept, and they were theoretical by-products of his own calculations in 1915.

Their presence is confirmed by two key phenomena: the gravitational influence on nearby stars; and the light generated by gas falling in. For instance, in the heart of our galaxy, huge stars orbit the supermassive black hole like planets in a star system. We know they must be orbiting something, though to our eyes there is nothing there. As for the gas, we can measure its speed via the Doppler shift in its light emissions as it spirals towards an unseen centre. These velocities are so extreme that only an object of truly staggering mass could be responsible, yet, again, there is no visible presence that an optical telescope might detect.

By the time the *Voyager* probes embarked on their journey in the 1970s, most astronomers were well on their way to being convinced of the existence of black holes thanks to these kinds of observations, despite how implausible they seemed. Yet overwhelming confirmation was not obtained until a whole century after Einstein's prediction; this came with the detection of gravitational waves in 2015.[1] And it was radio astronomers who unveiled, in 2019, the very first image of the shadow of a black hole.[2] This extraordinary achievement showcased the black hole at the heart of the giant, elliptical galaxy M87, followed in 2022 by an image of the black hole at the centre of our own Milky Way. These images, created using observations of radio waves emitted by material swirling right up to the event horizon, were assigned colours in the visible spectrum so we could 'see' the radio data.

The image looks like an orange circle, an uneven and lumpy ring with a conspicuous black absence at the centre. More than one internet voice pointed out a passing resemblance to the evil

Eye of Sauron atop his fortress in *The Lord of the Rings*. The analogy is a fitting one. Eight separate observatories had joined forces to form the Event Horizon Telescope (EHT), a fellowship of radio telescopes the size of the planet. Some were small, like hobbits, some loomed large, like wizards, but all played their part. Here were two stories of allegiances across borders and a battle between light and ultimate darkness.

Watching the live stream of the press conference unveiling the M87 image, I joined the rapturous applause that erupted from the audience. Even among my more stoic and reserved colleagues, there were a few tears. The gravitational wave detection two years before was indisputable and should have been enough for any scientist. But for all our scientific rigour, even for astronomers accustomed to the most abstract of data, 'seeing is believing' clearly still applies. Without radio telescopes, we would still be waiting for that visual confirmation, perhaps indefinitely.

Black holes are what's left over when the most massive stars collapse in on themselves. They are also an unwanted side effect of Albert Einstein's theory of general relativity. He never asked for them. He never particularly liked them. But he could not deny they were written into the fabric of his two theories concerning the nature of space and time: special and general relativity. Special relativity applies in scenarios where clocks or humans move at a constant velocity. General relativity, on the other hand, accounts for the effects of acceleration and provides an alternative interpretation of gravity. The mathematical model behind both theories fused the three dimensions of space with a fourth dimension, time, to form *spacetime*. We are used to moving about freely within three dimensions, but Einstein realised that we are also always moving through time too; we just don't get a choice in which direction. He showed that, in a sense, we all move through spacetime at a single universal speed: the speed of light. But we divide that motion differently between space and time depending on how fast we are moving. If you are moving around in the three space dimensions, you move more slowly in the time dimension. The startling result? The faster you move, the slower you appear to age, to an external observer.

We have observed this effect on a small scale by flying a clock on a supersonic jet around the world several times. When the clock is returned, it is running a few hundred nanoseconds slower than the clock that stayed on Earth. The pilot would also have aged a few hundred nanoseconds less than his colleagues, but the difference, of course, would not be noticeable. To see the relativistic effects of this *time dilation* on a human timescale, you have to be travelling through space close to the speed of light. But imagine we were able (technologically and ethically) to send one of two identical twin babies on a return trip to Alpha Centauri, one of the closest stars to Earth at 4.37 light-years, or 41.3 trillion km, away. If the rocket could travel at approximately 80 per cent of the speed of light, the twin sent into space would return with a biological age of six years old. Meanwhile, the twin left on Earth would have travelled faster through time and aged ten years.

'Twin paradox' thought experiments aside, the practical way to test Einstein's general theory of relativity was by testing the limits of the framework it was intended to replace: Newtonian gravity. Isaac Newton had outlined Newtonian gravity in 1687, describing an attractive force that acts at a distance between two masses. This theory assumes an instantaneous action, with no mechanism to explain how gravitational effects propagate, and treats space and time as distinct entities. Newton's model explains most interactions in the laboratory, even today, but there were a few discrepancies in how it predicted the movements of celestial bodies, in particular Mercury, that niggled at Einstein. Einstein once expressed that his single 'happiest thought' was the realisation that gravity was indistinguishable from acceleration and so would be best described by general relativity. Einstein found that if he thought of gravity not as a force, like Newton, but as a curvature of spacetime, any discrepancies disappeared.

Einstein's revised framework elegantly explains why objects with greater mass exert stronger gravitational forces. He proposed that massive objects distort spacetime more profoundly, creating stronger gravitational pulls and inducing greater accelerations in nearby objects. To visualise this, imagine an outstretched rubber sheet, which will represent the four-dimensional fabric of spacetime. A ping-pong ball placed on the sheet makes only a slight indentation, whereas a bowling ball creates a deep depression. Ball bearings placed on the sheet will naturally roll towards the deepest part of the dip, their motion accelerating as the slope steepens. If we roll a ball bearing at just the right speed and angle, and assume there is no friction or wind resistance to slow it down, that ball will fall into an orbit, just like a planet around a star.

Having a happy thought is one thing, but the task of mathematically articulating a 4D spacetime and its contents proved formidable even for Einstein, prompting him to seek the expertise of Swiss mathematician Marcel Grossmann. Grossmann's skills were instrumental in developing a robust framework that cohesively integrated space, time and gravity. With his theory now developed in a mathematical language, Einstein published four seminal papers in 1915, presenting a series of ten intricate and interconnected equations.

Einstein was happy with his theory, but there was something dark within it, a hidden consequence he would never fully come to terms with. It took a fresh set of eyes, those of German physicist Karl Schwarzschild, to take the theory to its limit. Schwarzschild wondered what would happen if he tried to solve Einstein's equations for a point source. We can think of this as crushing mass into an infinitely dense point in space (a cannonball on our rubber sheet, with the deformation stretching down to infinity). Schwarzschild showed that, if we were to approach that point, there would be a distance beyond which the gravitational force would be so large that even light could not escape. Schwarzschild had seen a consequence of Einstein's theory that no one else had, and what's more worked out a mathematical framework to describe it. More remarkable was that he did this while fighting for the German artillery, having enlisted in 1914. Sadly, as for so many others, the front line was to be Schwarzschild's own point of no return, and he died in 1916, but not before posting his work to Einstein. The radius he described is now known as the Schwarzschild radius, which defines where the event horizon lies around each black hole.

The term 'black hole' was coined decades after the concept was first proposed, and the story of its origin is as varied as it is fascinating. Many accounts credit American theoretical physicist John Wheeler with popularising the term at a 1967 conference. Frustrated by the unwieldy standard term, 'gravitationally fully collapsed object', Wheeler reportedly embraced the spontaneous suggestion of 'black hole' from a member of the audience, using it in his papers and talks. The phrase quickly caught on. If one digs deeper, it seems that either Wheeler (or the audience member) had already read an article by journalist Ann Ewing in 1964, entitled '"Black Holes" in Space'. If you look into how Ewing came up with the term, you will quickly wish you had not. Ewing reportedly heard American astronomer Robert Dicke relating the cosmological entity to a terrible incident whereby more than a hundred soldiers were imprisoned in 1756 in a dungeon the size of three double beds. This dungeon was called 'The Black Hole of Calcutta'. While some soldiers survived the night, most perished from suffocation and heat exhaustion – a chilling analogy for the concept of crushed mass.

Einstein was distinctly unnerved by Schwarzschild's findings, as were many others, seeing them as mathematical peculiarities, or an indication that something was wrong in general relativity. Over the decades, though, the proof mounted up, both for general relativity and for black holes. Optical astronomers observed stars seemingly orbiting an invisible mass in the centre of our galaxy, and then in many other galaxies too. Radio astronomers observed huge lobes of radio light emanating from galaxies, extending to such distances that there had to be a large central engine, something massive, but that they could not see directly. As our technology improved, black holes became less of a mathematical curiosity and more a reality to be chased and exposed. To prove general relativity was to legitimise black holes and, later, vice versa.

The greatest of proofs has to be the radio image of M87's black hole, or rather its silhouette. General relativity predicts that the intense gravity of a black hole bends light around it, forming a shadow. The shadow's shape and size depend on the black hole's mass and spin, and the EHT image reveals precisely the outline predicted. The term 'shadow of the black hole' was chosen carefully by the EHT team, as shadows are not a complete absence of light but rather regions where light is partially blocked. While a black hole is a three-dimensional object, the image we see shows only its two-dimensional outline. The immense gravity of a black hole pulls nearby gas into a swirling disc, wrapping around it like a whirlpool. This glowing material lies along our line of sight, meaning the central shadow is not entirely black anywhere. The brightest parts of the disc, closer to the event horizon, produce photons that are immediately deflected by the strong gravity and bent around the black hole, creating the bright ring seen in the image. Any photons that venture too close to the black hole are pulled beyond the event horizon, forming the shadow or, as Falcke described it: the 'gates of hell'.

While there are lumps and bumps caused by turbulent movement of the gas, the underlying ring is strikingly circular, rather than elliptical. According to general relativity, the warping of spacetime around a black hole should be symmetric, so any pronounced elongation would have signalled a breakdown in the theory. Instead, the image provides the strongest observational evidence yet for the existence of black holes, while also confirming that Einstein's equations remain valid under the extreme conditions found at the brink of a black hole.

Making an image of a black hole is not an easy feat. First of all, black holes are really good at hiding, veiling themselves in thick clouds of infalling gas and dust, blocking any visible light. Additionally, black holes are extremely small. This might seem an odd thing to say, as we routinely describe them as the most massive objects in space. But 'massive' in a scientific context relates only to an object's mass and not its physical size. Mathematically, black holes contain all their mass within a point of effectively zero volume: a singularity. But it is not this singularity which we observe; it is the last frontier of light surrounding it. When we refer to the size of a black hole, we are referring to its event horizon. The event horizon of the black hole at the centre of the Milky Way, for example, is only seventeen times the diameter of the Sun, or one-twelfth the size of Earth's orbit. At a distance of 256 quadrillion km, this makes observing the event horizon equivalent to the challenge of spotting a bagel on the Moon. At the short wavelengths of optical light, this resolution requires a telescope several kilometres wide: an engineering impossibility due to the perfectly smooth surface required of an optical lens. Radio light, however, is far more forgiving of bumpy, and even incomplete, surfaces. This means that we can piece together a planet-sized 'dish' using antennas strewn across the globe.

This method of observing is called 'very long baseline interferometry' (VLBI), where 'baseline' refers to the distance between the dishes involved, and 'interferometry' refers to the method of combining the observations from multiple antennas or dishes. Eight separate observatories took the first black hole measurements, stretching over distances of 11,000 km, from Mexico to Spain and from Hawai'i to the South Pole. These observatories have all been constructed at different times and for different reasons: the Atacama Large Millimeter/submillimeter Array in Chile is used to directly image exoplanets in formation; the South Pole Telescope is designed to study cosmic microwave background; the Large Millimeter Telescope in Mexico is designed to observe properties of the dust obscuring our observations across the galaxy. None were conceived to image a black hole. On its own, at peak performance, any one of these facilities could discern a bagel at a distance just shy of 400 km. Impressive, yes, but about a thousand times short of what is required. However, when these facilities are linked, they act as one gigantic dish, and the EHT achieves eyesight sufficient to read the newspaper headlines in New York from a cafe in Paris.

The EHT acts like one immense telescope, but it's no solid dish – just eight sets of antennas scattered across Earth, their signals woven together to mimic a complete dish riddled with continent-sized 'holes'. This renders the image blurry, yet there's sufficient information to sketch a rudimentary portrayal of a black hole. The analogy isn't perfect, but it's similar to attempting to paint a portrait while wearing a blindfold with eight little holes poked through. The outcome is not flawless, but it's remarkable given the resources available.

The telescopes needed to take their observations at the same time, and in roughly the same conditions. Though radio telescopes can deal with cloud very well, the accuracy to which these observations must be combined means that even small differences in how local cloud cover or rain affects the signal could wipe out the data from the black hole. There were many false starts where poor conditions at just one or two locations caused the planned observations to be cancelled. Once the telescopes managed to observe for a few hours over several nights, the scientists needed to painstakingly combine the data. But even getting all the data together was a logistical challenge. There was too much to upload and download, so the teams shipped their hard drives around the world. Because the data volume was so large, it wasn't easy to create or store back-ups, and so the astronomers were left to endure a tense wait to see if all the hard drives would arrive at the central analysis location. It was a long wait, too, because the scientists at the South Pole Telescope had to sit out the Antarctic winter before they could get to a post box.

Happily, all the hard drives did arrive safely, but if they had lost one along the way they may still have been able to make an image. The power of VLBI is that we can construct a picture out of very little data. For instance, during the second observation run in 2018, the team at the Large Millimeter Telescope in Mexico were held at gunpoint by a criminal gang. Although they emerged from the experience unscathed, they had been threatened with a return visit. Understandably, the scientists concluded that no data was worth risking lives. Besides, in the complex world of interferometry, where all antennas interconnect in a mesh, a broken link doesn't matter so much. Seven telescopes working in tandem are nearly as effective as eight; the only difference is a few missing fine-resolution features in the final image.

Imagine trying to find a window of perfect weather across observatories in Chile, Spain, Hawai'i, Arizona, Mexico, Antarctica, Greenland and France, all while diplomatically navigating the complex politics of rescheduling whoever was originally booked to use the telescopes. Now, picture assembling the world's foremost black hole photographers, in control rooms around the world, to aim their planet-sized camera, only to have their subject refuse to sit still. This is what happened when the EHT team attempted an observation of their primary target, Sagittarius A*, the black hole at the centre of the Milky Way. They likened the task to photographing 'a puppy chasing its tail'. Although they successfully captured the data, the relentless swirling of the gas around the black hole introduced significant motion blurring, to the point that they couldn't be sure what they were seeing. It was perhaps with a touch of frustration, but understandable pragmatism, that the team turned their attention northwards from Sagittarius to Virgo, a constellation housing the goddess of harvest and fertility. There, they found a more cooperative target: the supergiant elliptical galaxy M87.

Situated 55 million light-years beyond the Milky Way, M87 is an immense elliptical structure with trillions of stars and a central black hole with an event horizon 1,600 times larger than our own. So despite being 2,000 times further away, both black holes appear roughly the same size in the sky even though they are vastly different close-up. It is all a matter of perspective, just like how your thumb held up close to your eye can block out whole skyscrapers on the horizon. The behaviour of the black holes differs dramatically, however, betraying the difference in actual size. The enormous diameter of M87's black hole means the surrounding gas takes longer to orbit, so any features in the glowing ring move slowly over the few nights required for observations. By contrast, the gas around Sagittarius A* completes an orbit in mere minutes, flickering and shimmering with chaotic energy, a nightmare for any photographer, but especially one who expects to their subject to sit still for a single image taken over several days.

Given the choice between years of painstaking work to clean the Sagittarius A* data, or announcing the monumental achievement of imaging M87, the team chose the latter. I think some of us felt a pang of disappointment that humanity's first glimpse of a black hole wasn't of our own, but extracting a clear image required the development and testing of advanced software over several years. And in 2022, the radio astronomers' patient labour bore fruit: the EHT unveiled an image of Sagittarius A*'s event horizon.[3] Here, at last, was an answer to the question posed by Karl Jansky in his academic paper of 1935: just what was hissing at the centre of the Milky Way?

Later still, the EHT released images of both black holes in polarised light. Like iron filings tracing the magnetic field of a bar magnet, these images revealed the magnetic field lines around the event horizons. These findings underscored a profound truth: black holes are not merely destructive voids but whirring engines of cosmic creation. The spinning action of a black hole winds up magnetic field lines near its event horizon, much like a coiled spring. Eventually, the lines 'snap', releasing tremendous energy that accelerates any particles lying near the event horizon away at near-light speeds. These particles stream away in narrow jets that can extend millions of light-years. The particles themselves are invisible to optical telescopes, but the electrons emit bright radio light, as well as X-rays and gamma rays.

It is quite striking when one compares the optical image of a galaxy containing a black hole with an image taken by a radio telescope.[4] The optical image will show a galaxy with spiral arms of beautiful starlight, but no hint of the black hole that lies in the centre. Meanwhile the radio image has little hint of the galaxy itself, just two streams of radio light reaching in opposite directions, beginning pencil-thin before ballooning into wider splashes of light. We call the radio light emitted by those jets of particles 'radio lobes', and they do not appear in every galaxy – just 'radio-loud' galaxies, where the black hole is actively munching on nearby gas and stars. When we overlay two images of the same galaxy, one in optical and one in radio, we quickly see that the optical part of the galaxy is just one minor element of the complete image: the lobes can often cover more sky than the stellar part of the galaxy.

The galaxy Cygnus A, 700 million light-years away, hosts my favourite jets, because they remain needle-thin, almost to the point of invisibility, for a really long distance.[5] Then, suddenly, they puff out quite far from the black hole into lobes so brilliantly bright that they can be seen by any radio telescope that cares to look that way. The shape just perfectly portrays that speed and pressure, and so the power of the central black hole, and the whole structure reminds me of the steam that comes out of the ears of angry cartoon characters.

The EHT images of both M87 and Sagittarius A* were similar in shape, if very different in scale, demonstrating that the same mechanisms were at work in both black holes. Interestingly, though, radio images of M87 show a clear jet streaming out from the black hole,[6] while all observations around our own black hole suggest that Sagittarius A* has no jets to speak of. Both are spinning and both have magnetic fields, so what causes this difference? The answer is the fuel. M87 hosts an active black hole, one that has plenty of gas falling into it, and so plenty of material to chuck out when the magnetic fields snap. In fact, black holes are such efficient power plants that they need to consume less than one star the size of our Sun a year to be able to power these radio jets. Infalling material onto a rotating black hole can power jets with an astonishing 42 per cent energy efficiency – far greater than the Sun's own nuclear fusion, which manages just 1 per cent, itself 4 million times more efficient than a coal-fired power station. One might dream of a future where to switch on the lights means simply pouring some dust onto our household miniature black hole, harvesting the emitted energy and channelling it around the house. In reality, we've found recreating fusion tricky enough on Earth. Imagine trying to get planning permission for a local black hole: *Infinite spacetime distortion? Not in our neighbourhood!*

The absence of powerful radio jets around the mighty Sagittarius A* suggests it is on a near-starvation diet after billions of years of nibbling away at its fuel. There simply isn't much gas nearby for it to feast on, and what little material does exist is often too hot and diffuse to be easily captured. If it ever did erupt with jets, firstly they would be oriented out of the plane of the spiral arms, like a spinning top, far from Earth. If Earth did happen to get in the firing line, we would still be safe from immediate destruction due to the large distance but, over time, the intense radiation could gradually strip away our atmosphere and shower the planet with harmful energy.

Remarkably, black holes can be the architects of new stars while demolishing existing ones. As jets stream out of a galaxy, they encounter the interstellar and intergalactic medium: a dusty soup of atoms and molecules. Like a ship cutting through the sea, the jets become surrounded by eddies and whirlpools, creating a turbulent wake. Within the densest pockets of gas, the material can become dense enough to form stars: stars born of a black hole, from matter that was destined to be lost to the Universe for ever. Those particles find themselves creating light in the wake of ultimate darkness.

We began the chapter pondering what a *Voyager* probe would see on its way to the supermassive black hole at the centre of the Miky Way. In truth, they are going in the wrong direction entirely, but that does not mean they won't wander between smaller black holes, just as they wander between stars.

When we talk about black holes, we tend to focus on the

largest of them: the supermassive black holes that lie at the centre of galaxies. But the need for a description such as 'supermassive' is a clue to the variety of black holes out there. Most stars above twenty-five times the mass of the Sun will collapse under their own gravity when fusion stops, forming a black hole. These stars are not rare, and their lifetimes are relatively short at only a few million years, such that many black holes will have formed over the Milky Way's lifetime. Its single, central black hole, Sagittarius A*, is 'supermassive' in comparison to the *billions* of stellar-mass black holes that currently roam our galaxy. In case you are feeling disconcerted, we can detect these black holes using a variety of methods, mostly in the X-ray wavelengths, and we are in no danger: the nearest stellar mass black hole lies about 1,500 light-years away. Even though this is much closer than Sagittarius A*, which lies 27,000 light-years away, the event horizons of stellar mass black holes are so tiny that even the EHT cannot image their shadows.

If either *Voyager* were to get within that fatal Schwarzschild radius, and reach the event horizon, it would find itself torn apart, the intense gravitational force at its front end pulling it into a trail of atoms, in a process officially called 'spaghettification'. Once beyond that radius, even the last, desperate radio signal sent forth from its antenna would simply fall back in on itself. A silent, invisible and eternal ending.

Considering this destructive power, it is remarkable that we know anything at all about black holes. Yet physics and mathematics have proved to be powerful tools, allowing us to study them from a very safe distance. Over the century since they were first conceived within the equations of Einstein's general relativity, the evidence for their existence and prevalence has piled high. Still, we are such visual creatures that many, I suspect, held on to doubts – doubts only a picture could truly silence. Only radio astronomy, implemented by teams of astronomers spread across continents, was able to provide that irrefutable image of a black hole that we all secretly craved. Even for the most logical scientist, at home in the abstract frameworks of spacetime and mathematics, a picture is still worth a thousand equations.

9

NEUTRON STARS

> A cubic centimetre of this star would weigh as much as the whole of humanity.
>
> Jocelyn Bell Burnell,
> 'My Century', August 1999

Part II of our journey has seen us spiral along the arms of the Milky Way, before halting to stare directly into its dazzling heart. Dizzy from the motion, and with our eyes still blinking away the glare of Sagittarius A*, you'll be glad to know our next destination requires just a small turn of the head. When we are facing the galactic centre from Earth, we see the dense brilliance of the constellation Sagittarius. But if we look a little to the left, by about sixty degrees, we will spot a lesser-known constellation: Vulpecula.

Despite still sitting within the ribbon of the Milky Way,

Vulpecula is set against a far darker backdrop. This is because we are not gazing into the densely packed central bulge of the galaxy any more, but just to one side, through its outskirts. Vulpecula, the constellation of the Fox, is small, faint and easily overlooked, barely visible in light-polluted skies. Despite this, it is one of the easiest constellations to locate, because this elusive fox is nestled in the middle of one of the most recognisable shapes in the night sky, the 'Summer Triangle'. The stars outlining the triangle, Vega, Deneb and Altair, far outshine any of Vulpecula's members, at least in visible light. But in the radio, the balance of power shifts.

Foxes are, in many mythologies, sly tricksters, and this little fox lives up to its reputation. Hidden in its den is a star we cannot see with the naked eye, or even with the artificial eye of a space telescope, yet it howls across the galaxy in radio light. This is no ordinary star, but a neutron star: the dense, collapsed core of a once-massive Sun, its fuel spent and its compression complete, now reduced to an inanimate sphere about the size of a city.

For radio astronomers, these stars are a revelation. For physicists, they are a dream come true. These stellar remnants are heated to millions of degrees and compressed to densities beyond imagination, placing neutron stars at the crossroads of astronomy and quantum physics. Inside these celestial laboratories, the largest scales of the Universe meet the smallest. It is enough to tempt even the most dedicated condensed matter physicist to swap their microscope for a radio telescope.

In the summer of 1967, a young graduate student named Jocelyn Bell Burnell was poring over reams of paper spilling from a chart recorder, a mechanical device that traced radio signal strength with an ink pen, line by line, onto a moving strip of paper, much like the output from a heart monitor. Among the familiar murmurings of the Sun, planets and Milky Way, something curious stood out: a series of pulses that repeated every 1.337 seconds, with unwavering precision. This 'scruff', as she called it, defied explanation.

The telescope was the brainchild of trailblazing astronomer Antony Hewish, Bell Burnell's PhD supervisor at the University of Cambridge.[1] Hewish was looking for the bright radio emission from radio jets around supermassive black holes, a phenomenon known as a quasar, but he needed an enormous telescope to do so. He did not have the funding for a massive dish like the 305-m one at Arecibo Observatory, so he cheated. Or, rather, he used the technique of radio interferometry and designed an array that mimicked the power of a single gigantic telescope by combining signals from many antennas, just as the EHT would go on to do on a planetary scale. Over two years, Bell Burnell and her fellow graduate students constructed the telescope by hand in a field a few kilometres southwest of the university. They hammered more than a thousand wooden posts into the ground, threading them together with some 190 km of wire and cable.* The result, resembling row upon row of washing lines, stretched across an area equivalent to fifty-seven tennis courts.[2]

Unlike Jansky's 'merry-go-round', which rotated to scan the sky, this telescope remained stationary, watching as the heavens drifted above. The signals from the telescope's extensive array of wires were combined in such a way that it could focus in different directions, revisiting the same patch of sky every four days. Each day, the chart recorder spat out 30 m of chart paper and, each day, Bell Burnell followed the lines of ink on every sheet, looking for anything unusual. It was not immediately obvious, but she

* It was good exercise. When a fair came to town during that period, one of the students swung a hammer so hard they broke the strength-test.

soon noticed that every 120 m of paper, a strange pattern of pulses repeated.

Initially, the team speculated that the signal might be synthetic: perhaps the Russians had launched a new satellite? That idea was soon dismissed, as, if the signal had come from something orbiting Earth, it would have varied in timing and strength as it flew overhead. It took four days, equivalent to 120 m of chart paper, for the telescope to observe the same position in the sky. In this case, it seemed that every time the Vulpecula constellation climbed overhead, the pen briefly scribbled up and down every 1.337 seconds. The pulses were faint, almost imperceptible among the noise of Earth-based interference and celestial radio emissions. That Bell Burnell noticed such tiny signals repeated across football-pitch lengths of chart paper, when she was looking for something entirely different, is extraordinary. It is the perfect example of the deep familiarity radio astronomers of the 1950s and 1960s were expected to have with their instruments, and the radio skies they studied.

It was the regularity of the bursts that was so unusual. The radio noise from the centre of the galaxy was a constant hiss, and stars and planets had never been observed emitting pulsing radiation in that way. But there was no doubt: something within Vulpecula was generating regular bursts of radio noise, as if following a metronome. There was no 'optical counterpart', nothing but the signal itself, to help them work out what this was.

The first step came when Hewish and Bell Burnell realised they could estimate the size of the pulsing radio source. If, hypothetically, the Sun could switch on and off instantly, we would not see its light change all at once. The centre of the disc, being the closest point to us on the sphere, would appear to light up first, and the edges, being 700,000 km further away, would light up a few seconds later. In this way, the length of the pulse tells us the size of the star, or other luminous object – the longer the pulse, the larger the source of light.

Bell Burnell had detected clusters of pulses every 1.337 seconds, with each individual pulse lasting only ten milliseconds. Following these physical arguments, they calculated that the source had to be less than 5,000 km in diameter: more like the size of a small planet than any star they knew. Yet no planet behaved in this way. What could it be?

Having entertained and dismissed the idea that the Russians might somehow be involved, they began to wonder if it could be something a little more . . . extraterrestrial. Hewish, Bell Burnell and colleagues half-jokingly labelled the mysterious repeating scruff 'LGM' on the charts, for Little Green Men. The precision of the pulses was uncanny, so much so they could not quite believe it was natural. And now they knew the signal was coming from something smaller than Earth. What other explanation could there be than an alien radio beacon trying to make contact?

Now the astronomers knew what they were looking for, they pored over the charts and discovered three more bits of repeating scruff. One was not far from the first, still in Vulpecula in fact, and the other two were in the constellations of Cetus and Cassiopeia. The idea of four separate groups of Little Green Men, all producing radio signals at the same frequency from different parts of the sky, was clearly implausible, laying LGM to rest.

The group were now sure that they had discovered a new type of astrophysical source, so they went in search of exotic theoretical bodies that might explain the data.[3] Thirty years earlier, in 1934, astronomers Walter Baade and Fritz Zwicky had hypothesised that the collapsing core of a dying star could become so dense that protons and electrons would combine, leaving behind an optically invisible, compact object composed almost entirely of neutrons – then a newly discovered subatomic particle. After over thirty years of this theory effectively sitting on the shelf, neutron stars had received a surprise experimental confirmation. As silent in the optical as a shy backing singer with a sore throat, in the radio, some neutron stars have a baritone to blow your socks off.

The term 'neutron star' combines two very different scales: the minuscule world of subatomic particles and the vastness of stellar phenomena. But at their core, quite literally, stars are only ever lucky clouds of gas, granted a brief respite while the crushing force of gravity is held back by the pressure of the tiny nuclear reactions within. They are always doomed, eventually, to collapse in on themselves.

A star in its prime, like our Sun, is actively fusing hydrogen into helium, releasing the light essential for our survival, while creating an outward radiation pressure. When nuclear fusion processes cease, that outward pressure diminishes, allowing gravity to prevail. For truly massive stars, the gravitational force is so strong that the star collapses promptly into a black hole, as we saw in the last chapter. But for less massive stars, another kind of immortality awaits.

When a star up to about ten solar masses begins to collapse, it sheds the vast majority of its mass in a spectacular supernova explosion. Only the core remains: a much smaller, denser fragment which continues to compress under its own immense weight. The atoms in this core are squeezed into a tighter and tighter space, until the electrons within it are forced so close that they generate a powerful new repulsive force. This force, called *electron degeneracy pressure*, arises from a fundamental principle of quantum mechanics: electrons do not like being in the same place at the same time, and will do anything to resist that situation. Unlike in nuclear fusion, where a continual supply of fuel is required, electron degeneracy pressure is unwavering. This outward pressure balances the inward pull of gravity, halting the collapse indefinitely. This category of remnants, only about the size of Earth yet holding as much mass as the Sun, are known as white dwarfs.

Approximately 97 per cent of the Milky Way's stars are destined to become white dwarfs, including our own Sun, but it is not a bad way to go. These dense stellar remnants have an intense gravitational field that affects the material within them. Over time, heavier elements sink towards the core and lighter ones rise to the surface, forming an atmosphere primarily of hydrogen and helium. Over billions of years, the white dwarf slowly radiates heat and cools, causing the carbon at its surface to crystallise, effectively turning it into a gigantic diamond. For a long while, far longer than their lifetimes as a working star, far longer than the lifetime of the Universe so far, they will still be visible due to the thermal radiation resulting from their 'white' hot temperatures, like the coals of a spent fire. Eventually, though, after quadrillions of years, they will cool such that their thermal radiation shifts out of the visible, and those diamonds will finally fail to shine on, entering eternal obscurity as a black dwarf.

But what happens for stars above ten solar masses, stars so massive that the gravitational pressure of their collapsing layers can overcome even the electron degeneracy pressure? Atoms, whether oxygen, carbon or einsteinium, consist of a nucleus made up of positively charged protons and neutrally charged neutrons, all surrounded by a swarm of negatively charged electrons. Under sufficient pressure, protons can 'capture' electrons, their opposite charges cancelling, and convert themselves into neutrons. Shortly after the neutron was experimentally confirmed in 1932, building on the identification of the electron in 1897 and the proton in 1909, Baade and Zwicky suggested that this electron capture process could occur within the high-pressure, high-density environment of a large collapsing star.

In these stars, the core is placed under such immense pressure from the surrounding layers of matter that the atoms are shattered, their electrons and protons combining into a sea of neutrons. Gravity's formidable force can compress a star many times the mass of our Sun into a sphere roughly the size of Manhattan in just a few seconds, and the core becomes exclusively neutron-based, with no room for further compression. Like a human trying to squeeze a block of lead, gravity finds itself meeting resistance, halting the collapse and marking the birth of a neutron star, with a volume around 250 million times smaller than Earth.

As with white dwarfs, there is no fusion process occurring within neutron stars. But, while white dwarfs can still be seen by their thermal emission in the optical part of the electromagnetic spectrum, neutron stars are far hotter and emit thermal radiation at much shorter wavelengths. Even then, their diminutive size means that they emit far less light overall, making them faint or invisible at every wavelength. Unlike the glamorous, diamond-encrusted fates of their white dwarf siblings, they remain a silent lump of neutrons floating in space, a convergence of micro- and macroscopic worlds.

Whereas white dwarfs are able to retain an atmosphere of lightweight gases, neutron stars have such strong gravitational fields that even these sparse gases are pulled inexorably inwards. As the core of a massive star collapses, the remaining material completes its fusion journey: lighter elements like hydrogen and helium are forced together and converted into heavier elements. These elements form the metal jacket that makes up the neutron star's dense, iron-rich crust. While living stars are nebulous, gaseous entities with no clear surface, in death, neutron stars become more solid, more like a planet than a star. Gravity smooths their surface to millimetre precision, creating a structure as flawless as polished marble. Below, there is a sea of free but compacted neutrons, a soup of broken atoms.

Neutron stars are, barring black holes, the most dense and high-pressure environments in the Universe. There are numerous fun facts about neutron stars (as physicists, we have a slightly different idea of what constitutes 'fun'). For instance, to mimic the density of a neutron star, one would need to compress the entire human population into a space the size of a sugar cube. As astronomer Frank Shu remarked: 'How much of humanity is empty space!' Or we would need to collapse the entire mass of the Earth inside a two-storey house. You had better hope it wasn't your house, for if you were to come into contact with a neutron star it would wear you like a coat. Its immense gravitational pull would instantly convert all your atoms into neutrons and spread you like butter into a thin layer covering its surface. See, fun!

Studying a neutron star up close is therefore simply not an option, but it's not easy to study them from a distance, either. While the supernova is very hot and bright, without ongoing fusion they fade in the optical wavelengths over a matter of weeks. However, there is a particular type of neutron star that shines so brightly in the radio that we find them even when we're not looking.

When Zwicky and Baade developed their neutron star theory in the early 1930s, they realised these objects would be invisible in the optical. At the time, optical astronomy was all of astronomy, and there was no reason to make a concerted effort to look for something expected to be entirely invisible, however intriguing the idea. By the time Bell Burnell was hammering wooden posts into a muddy field in the 1960s, radio astronomy had been established as a mainstream science, and stars, planets and galaxies were no longer thought to be silent at the longest wavelengths. Perhaps neutron stars weren't entirely invisible after all.

With a much greater understanding of how celestial radio emission worked, the team realised that if neutron stars were emitting radio waves from just one part of their surface, we would see a flash of radio light with a regularity matching the star's rotation. The Cambridge team had four astrophysical sources which repeated every 1.337 seconds, 1.275 seconds, 1.187 seconds and 0.253 seconds. As a star collapses, any rotation it has will be magnified, like an ice skater bringing in her arms to increase the speed of her spin. If neutron stars did exist, they would be able to spin at extraordinary rates, because their collapse had been almost total. There would be stars spinning every second, several times a second, hundreds of times a second even, easily fitting the Cambridge team's data. Hewish and Bell Burnell were now convinced they had confirmed the existence of neutron stars – specifically, a unique kind of neutron star that beams its otherwise invisible presence across the cosmos through radio waves: a *pulsar*.

The strange name comes from the term '*pulsa*ting *r*adio source'. As the neutron star rotates, its magnetic field is dragged round with it. Particles within that magnetic field are accelerated, causing the emission of electromagnetic radiation that is then funnelled along the magnetic field lines to produce focused beams of radio light emanating from each pole. If the poles of the neutron star are oriented such that the beam sweeps over Earth, then we will see a pulse of electromagnetic radiation. If not, it is doomed to obscurity as both optically invisible and radio-silent. Because the neutron star rotates at a constant rate, these pulses are seen with a constant period, just like a cosmic lighthouse.

While all pulsars are neutron stars, not all neutron stars are pulsars: that distinction is awarded based on whether it is observable in the radio wavelengths. Pulsars are just neutron stars with something to say. That being said, the vast majority of neutron stars we know about, we only know about *because* they are pulsars. While it is not impossible to spot one with modern technology and multi-wavelength observatories, it is far trickier to detect neutron stars without that pulsing behaviour.

★

The radio telescope built by Jocelyn Bell Burnell was simple by modern standards, and its data was overwhelmed with radio signals from human communications systems, the environment, even its own electronic components. Now, with digital signal processing, the heartbeat of a pulsar leaps out from the data with stunning clarity.

We now know there is huge variety in the patterns of radio light emitted by pulsars. Some beat faster than others, some slower, and some even skip the odd beat. Millisecond pulsars spin hundreds of times a second with a reliability rivalling an atomic clock, making them useful for testing grand theories of relativity out in the Universe, as we will see in the next chapter. But even the pulsars with an arrythmia can tell us something interesting about the inside of a neutron star, an otherwise impossible environment to explore.

In these cases, the pulsar's heartbeat begins to race, temporarily speeding up in an event called a 'glitch'. It is the mark of a seismic episode, where the crust of a pulsar buckles under the stress induced by the star's own rotation. Just as an earthquake tells us something about Earth's tectonic plates and the mantle beneath, a pulsar glitch provides a rare opportunity to probe the inner workings of the secretive neutron star. The frequency, duration and effects of these glitches inform the 'equations of state' studied by physicists, which describe how matter behaves under extreme conditions.[4]

Inside the neutron star, the neutrons are *superfluid*, a peculiar kind of liquid with zero resistance to flow. Whereas we might normally expect solids to form under high pressure, as in the Earth's inner core, the extreme pressure in a neutron star corrupts particles and prevents them forming a stable lattice. If set swirling, as in a rotating neutron star, a superfluid creates whirlpools that don't lose energy as they flow. In comparison, the iron jacket of the pulsar does lose energy as it moves and so slows down ever so slightly over time. During a glitch, the mismatched speeds of the superfluid and the crust suddenly correct, such that the crust speeds up. On Earth, superfluids are studied for applications ranging from quantum computing to spaceship accelerometers. But even our best labs can't replicate the extreme conditions of a neutron star: its density, magnetic fields or sheer size. A neutron star's 10-km diameter may be tiny for a star, but as a natural laboratory, it is colossal.

Neutron stars may also host quark-gluon plasma, a theoretical state of matter where protons and neutrons are dismantled into their building blocks: quarks and gluons. This 'particle soup' might exist in the core of a neutron star, as the wreckage left after the collapse shattered the star's atoms. Particle accelerators on Earth, like those at CERN, can recreate the extreme densities and magnetic fields in a quark-gluon plasma, but these experiments also generate immense heat, which differs vastly from the cold, dense environment of a neutron star. 'Cold dense matter' remains a mystery, one we hope to understand more fully by studying neutron stars.

When Jocelyn Bell Burnell first noticed that bit of scruff on her charts, she brought together two seemingly unconnected worlds: the microscopic world of atomic physics and the gigantic scales of space, all using radio astronomy. Sadly, Bell Burnell is less known for her scientific breakthrough than for the controversy that surrounds it.

In 1968, Hewish and Bell Burnell published a paper on their discovery, and it became front-page news not only in the astronomy community, but in the wider world too. If the discovery of a new kind of star wasn't enough to excite you, there was the sensationalism of the Little Green Men and, at the centre of it all, there was something even more alien . . . a woman! Seeing a woman in the sciences was rare in the 1960s and Bell Burnell became the face of the discovery. Her picture appeared in national newspapers and one tabloid even printed a headline along the lines of 'Girl Makes Discovery' alongside her estimated body measurements. She was subject to inane lines of questioning: whether she was the same height as Princess Margaret and how many boyfriends she typically had at once. A photographer even implored her: 'Look happy dear! You've just made a discovery!'[5] When the press interest had died down, Bell Burnell could continue her PhD which, if you remember, was meant to be on quasars. In her final thesis, she relegated pulsars, unbelievably, to an appendix.

In 2018 she won the Breakthrough Prize of £2.3 million and gifted it all to the Institute of Physics to fund women, under-represented ethnic-minority and refugee students to become physics researchers.[6] Do you know what she didn't win? The Nobel Prize. Antony Hewish did, though, in 1974, for that first pulsar discovery. Back then there wasn't a precedent of students being rewarded with such a high honour, instead it was those that had designed the experiment and led the research, so the slight was not out of place for the times, however hard it is to understand now. To her credit, Bell Burnell has never appeared bitter about this, despite her pivotal role in the breakthrough. There are plenty of scientists willing to be bitter on her behalf, however, referring to it as the No-Bell Prize.

And what happened to that humble radio array that enabled pulsar science to begin? Well, I can tell you first hand. I was visiting another telescope in Cambridge when our guide casually mentioned that we were driving past the original pulsar telescope, hidden behind a row of bushes.[7] I believe I made some sort of involuntary yelp of surprise that was enough to stop the car. I leapt out and hot-footed it alone through an overgrown, prickly hedge.

And there it was.

The copper wire that once stretched between the posts has largely all been stolen, but the weathered wooden posts still stand, arranged just where Jocelyn Bell Burnell hammered them into the ground. In the telescope's prime, the grass was kept short using sheep, to prevent electromagnetic signals from a lawnmower interfering with the experiment. Now retired, the array sits quietly in a field of overgrown grass, growing old behind a hedge, with not even sheep for company.

Every now and then, though, Vulpecula the Fox drifts overhead, a visitor to an old friend. And deep within that den of stars, the first neutron star ever detected pulses on and on.

10

GRAVITATIONAL WAVES

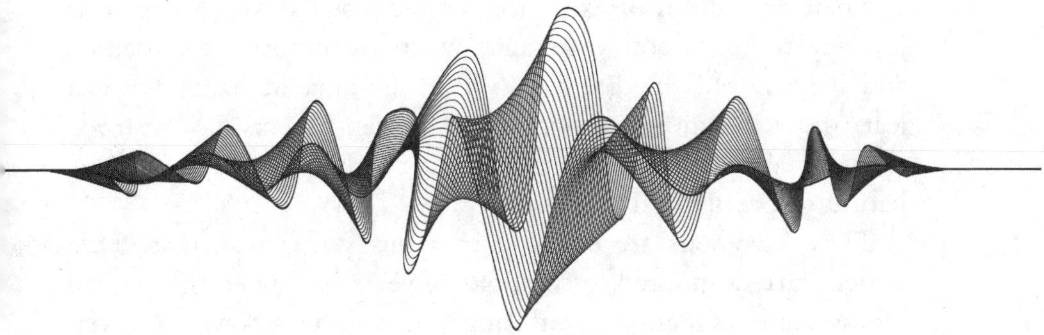

> I want you to feel it, in your bones, in your breath, when two black holes colliding a billion light years away sends a tremor through spacetime that makes every cell in your body stretch, and strain.
>
> Katie Mack, 'Disorientation'

Twenty-one thousand light-years away, in the constellation of Aquila, a pair of stars are locked in a dance to the death. You may have even glanced in their direction without knowing – perhaps when admiring the night sky, a glass of wine in hand. Aquila, bordered by Sagittarius and Aquarius, is shaped by a scattering of bright stars that sketch the shape of a great eagle, wings outstretched as it soars through the night.

Tucked above the eagle's right wing lie the twin stars of the Hulse–Taylor binary system. But you won't spot them – they're invisible to the naked eye, and even to *Gaia*, the great cartographer of the stars, whose precision drops dramatically beyond 10,000 light-years. These two lie more than twice that distance, and are extremely faint to begin with, far beyond our reach in visible light. In the days when the constellations were first drawn and mythologies mapped to the stars, this duo were entirely hidden.

Their only proof of existence is a radio signal which began its journey to us 21,000 years ago, when mammoths still roamed and ice covered the Earth. With an antenna in hand, we can learn how the movements of these unseen objects have warped the very fabric of the Universe, offering a glimpse into the strange nature of gravity and spacetime.

These two stars are engaged in a turf war, an ancient battle which started hundreds of millions of years ago when they began life as young supergiants orbiting one another. Now, however, they are neutron stars, mere shadows of their former selves, stripped of fuel and weight. They are minuscule: if the Moon were the size of a car, a neutron star would be the size of a marble. Don't let their diminutive size fool you, though. They both possess the agility of a featherweight and the strength of a heavyweight. The stars are balanced adversaries in death, as they were in life, circling each other like boxers in a ring. It is a draining duel, a gravitational slugfest, serving only to delay their inevitable and mutual destruction.

Neutron stars on their own are not rare and, since Jocelyn Bell Burnell's discovery in the 1960s, we have detected thousands using radio telescopes. But a binary system consisting of stellar remnants is much rarer, despite the fact that 85 per cent of the stars we observe start off in systems with more than one star. Our Solar System is rare in this respect, and thankfully so, as models show us it is far harder to create a stable, habitable planet when there are two Suns orbiting each other to contend with. The supernova preceding the neutron star formation is violent and highly disruptive, such that even though most stars start with at least one partner, they are eventually likely to be obliterated or kicked out of orbit. Of all the pulsars we have observed, about 95 per cent of them are alone – far too many for them only to have formed from rare, single star systems like the Solar System.

Many binary systems don't survive the death of one of their members but, even if they do, they then have to survive the death of the second star in order to stay together. Survival relies on that star dying with grace: gently casting off its outer layers and becoming a white dwarf, rather than exploding and becoming a neutron star. Which is what makes the binary system in the constellation of Aquila so extraordinary: it contains not one, but two neutron stars, implying it endured two apocalyptic supernovae. This is staggeringly rare. Fewer than 1 per cent of all observed neutron stars are paired with another neutron star. They are an exception to the rule, extraordinary survivors of a process which left many of their siblings facing an eternity alone.

This pair did survive these pre-match training rounds however, even as the force of the supernovae flung them through the galaxy at several hundred kilometres per second. Yet still they remain connected, like a pair of tussling fighters spilling out of the ring and into the streets. Due to their distance from us, they have appeared to move only by a star's width within Aquila in the decades we have been watching them with radio telescopes.

If both stars in the pair were just ordinary neutron stars, we'd likely have no idea they were there at all. After the brilliance of their supernovae, the stellar remnants themselves cool and fade out of the optical rapidly, sliding into invisibility. And indeed, one half of the Hulse–Taylor binary has done just that: it is a neutron star, and nothing more. It spins in silence, giving no hint of its presence in any wavelength of light.

Fortunately, that invisible neutron star's companion is a pulsar, which is not quite so willing to fade gently into the darkness. In this star's case, strong magnetic fields are sweeping up high-energy particles and funnelling them to the magnetic poles, where they combine to create bright beams of radio light. That's what makes this system visible to us: through a fluke of orientation, that beam of radio light sweeps over Earth again and again as the neutron star spins. The presence of a silent partner is imprinted on the pulsar's radio signal: it's gravitational pull creates a small, rhythmic wobble in the pulsar's motion, subtly altering the timing and pattern of the radio pulses we receive.

It is quite possible, likely even, that both neutron stars are in fact acting in this way, but only one is oriented such that Earth lies in the path of that sweeping radio beam. Another planet, with a different viewing angle, may well see the neutron stars swap roles, with our pulsar falling silent to those alien observers.

In the previous chapter, we looked at the ways in which radio astronomers and condensed matter physicists use the heartbeat of radio-loud pulsars to 'look' inside neutron stars and investigate the most extreme forms of matter. The Hulse–Taylor binary system in Aquila ups the stakes, offering us an even rarer, even more extreme environment in which to test Einstein's theory of general relativity.[1] These dense entities are separated by a distance so small that the system could comfortably fit inside our Sun. The intense gravitational field means they spin around each other so fast that they complete an orbit in only seven and three-quarter hours, creating ripples through spacetime, like a speedboat doing a doughnut in a placid sea. The 'gravitational waves' from this system are not detectable directly, but their emission subtly alters the pattern of loud pulses of radio light which are very much observable. Pulsars are among the most studied objects in astronomy because of their ability to significantly ruffle spacetime *and* let us know about it.

The Hulse–Taylor confirmation of these spacetime ripples relied not on sophisticated gravitational wave detectors but on two researchers sat in the control room of an ageing radio telescope in 1974.

Einstein's theory of gravity is far more intricate than Newton's, which elegantly summarised gravitational force as increasing with mass and decreasing with distance in one simple equation. For 350 years, Newtonian physics provided an excellent framework for understanding phenomena from falling apples to the Moon's orbit. However, as technology advanced, scientists began to notice deviations when dealing with extreme masses or accelerations.

A key difference between the two theories lies in how gravity operates. For Newton, gravity was instantaneous – remove the Sun, and Earth would immediately veer off into outer space. Einstein proposed instead that changes in gravity travel as gravitational waves, much as light waves carry electromagnetic energy. These ripples in spacetime travel at the speed of light – fast, but not instantaneous. According to Einstein, if the Sun disappeared, it would take eight minutes both for our sky to turn dark and for Earth to stop orbiting the empty space.

Crucially, just like light, gravitational waves carry away energy from a system. In a Newtonian Universe two stars could circle each other indefinitely, never losing energy. In Einstein's general relativity, the stars would lose energy as they produced ripples in spacetime, a kind of gravitational friction. In Einstein's Universe, orbiting bodies gradually lose energy and spiral closer, unlike the eternal orbits predicted by Newton.

Gravitational waves are produced by any moving mass, but their intensity depends on the masses and accelerations involved. Your body, for instance, causes minuscule distortions in spacetime, and your movements generate gravitational waves. But these waves are far too faint to detect. Even planets fail to produce a measurable splash.

As the Earth orbits the Sun, it does create gravitational waves, resulting in a minuscule slowing and shrinking of our orbit, by about a trillionth of a metre per year. This effect is so minor that it will remain insignificant over the entire lifespan of the Solar System. To observe a measurable change in orbits on a human timescale, we need exceptionally massive, rapidly moving objects, which emit more gravitational waves. In the constellation Aquila, we find an ideal example: a binary pair of neutron stars. These stellar remnants, among the most massive objects in the Universe, whip around each other with extraordinary acceleration, producing gravitational waves strong enough to observe indirectly.

For forty thousand years the Hulse–Taylor stars have been locked in battle, circling each other over and over again. According to Newton's theory of gravity, their separation should remain constant. However, general relativity predicts that over each 7.75-hour orbit, the system loses enough energy for the stars to edge about 3 mm closer. This tiny shift shortens their orbital period by about 77 microseconds (0.000077 seconds) per year. Surely a change so minute is impossible to measure in an astronomical system light-years away? Well, not if one of the stars happens to be a pulsar, producing flashes of radio light with the regularity of a ticking clock.

The Hulse–Taylor binary system was discovered in 1974 from the control room of Puerto Rico's Arecibo Observatory.[2] Radio astronomer Joseph Hooton Taylor and his PhD student Russell Alan Hulse were actively looking for pulsars, following their discovery by Jocelyn Bell Burnell seven years earlier. Each time Aquila the Eagle flew into view of the telescope, they spotted a piece of repeating scruff amongst the radio data. They were looking at a neutron star spinning seventeen times per second, equivalent to a rotational period of just fifty-nine milliseconds.

Since Bell Burnell's pioneering work, pulsars had become key targets for radio astronomers, and by 1974 over a hundred pulsars had been identified across the sky.* The rotation rates varied widely, from 4 seconds to an astonishing 1.5 milliseconds. While the speed of the Hulse–Taylor pulsar wasn't exceptional, Taylor and Hulse noticed something unusual. The pulses, arriving every fifty-nine milliseconds on average, shifted subtly: arriving slightly earlier for nearly four hours, then slightly later for the next four. Pulsars are so accurate that the astronomers could not see how its rotation would vary in such a structured way. Even pulsar glitches are sudden, and not repetitive.

The answer lay in a binary system. As the pulsar orbits its partner, its distance from Earth fluctuates, causing the radio pulses to travel slightly further or shorter distances. The variation they were seeing was the signature of a 7.75-hour orbit. We cannot see that partner, but we can model the orbital mechanics from the pulsar timings and work out that it has a mass best fit by a neutron star. This was the first binary system ever found to include a pulsar, and Hulse and Taylor eventually won the Nobel Prize for Physics in 1993 for 'the discovery of a new type of pulsar, a discovery that has opened up new possibilities for the study of gravitation'. Within hours of detecting that inaugural pulse in 1974, Hulse and Taylor grasped the system's potential significance. They had uncovered an extraordinary system, one intense enough to produce significant gravitational radiation and equipped with a pulsar acting as a precise cosmic clock. This allowed them to measure the orbital period of the stars and detect any changes in their separation.

They had the means to provide the first definitive evidence of the existence of gravitational waves, if only they could show that the binary's orbit was shrinking with time. The predicted change

* Several thousand are known today.

in that 7.75-hour period was tiny, and so the precision of their instrument meant they needed extensive data, likely collected over several years, in order to be sure. Embracing the patience required in pulsar science, they settled into their work amid the Puerto Rican jungle's chorus of tiny tree frogs and vigilantly observed the Eagle as it soared into view again and again, singing out its secrets.

Hulse and Taylor knew what they had early on, everyone did, but they were forced to just keep listening, building up the data and pushing down the errors. Imagine someone shouting to you across a crowded cafe, from so far away that you can barely hear them over the general hubbub. If they shout once, their message is pretty much lost, apart from a word or two. But if they shout the same thing again you might hear another snippet, in between the conversations going on around you. Once they have shouted it ten times, they are probably getting some annoyed looks, but you are also likely to have deciphered most of the message, even if there are some mistakes, or gaps. The principle is the same with a faint, repeating radio signal: the more times you hear it, the more confident you become in decoding its meaning.

Taylor's vigil lasted seven years. After the first three, Hulse stepped away from the project, perhaps feeling that their confirmation of gravitational waves was now inevitable. Taylor then partnered with a new postdoctoral researcher, Joel Weisberg, and together they published the long-awaited confirmation in 1982. Their meticulous observations revealed that the neutron stars were not fixed in their orbits, as Newtonian physics might suggest, but spiralling closer together due to the energy lost as gravitational waves.[3] The pulsar's slowing period matched Einstein's predictions with striking precision: seventy-seven-millionths of a second per year.

Weisberg and Taylor never did look away, in fact. In academia, their collaboration became a quiet tradition, with papers emerging every few years authored by 'Taylor and Weisberg' or 'Weisberg and Taylor', echoing the twirling motion of the binary stars they continued to observe. Over four decades, they monitored their celestial companions using Arecibo and other radio telescopes, creating one of the most precise, long-term measurements of gravitational radiation ever achieved. In thirty years, the stars have drawn just 100 m closer, yet the message is undeniable: as objects move through space, they leave ripples of gravitational waves in their wake.[4]

A direct detection of gravitational waves only came along in 2015, over four decades after Hulse and Taylor measured their presence indirectly from the Aquila pulsar binary. Gravitational waves do not simply lift and set down objects as a wave might rock a boat. The boat is separate from the water, but all objects are part of spacetime and so are stretched and distorted as the wave passes through. Everything, including you, is constantly being stretched and squashed by these waves, though by amounts so infinitesimally small they pass unnoticed. To directly detect gravitational waves, we needed to catch them in the act of stretching and compressing spacetime, a feat requiring a cataclysmic event like the collision of black holes or neutron stars. Such events produce a tsunami of gravitational waves, yet even these immense ripples barely distort Earth, necessitating the precision of twenty-first-century technology.

The Laser Interferometer Gravitational-wave Observatory (LIGO) is Earth's most sensitive gravitational wave detector. It consists of two identical observatories: one in Livingston, Louisiana, and one in Hanford, Washington. On each site, there are two 4-km tunnels arranged at right angles. A laser at their junction splits its beam, sending light to the ends of each tunnel, where it is reflected back. If spacetime is calm, the troughs of one beam will meet with the peaks of the other, and the waves will cancel out. If a gravitational wave is passing through, however, one tunnel will be stretched slightly longer than the other, before the stretch–squash action is reversed. The difference in the tunnel length means that one beam of light has travelled further, and the troughs and peaks will be out of sync with the other beam. This creates a stripy pattern of light known as an 'interference pattern', when the beams are recombined at the junction.

The LIGO set-up is so precise that it can detect weather patterns on the opposite site of the continent, surges of water in the ocean, and every earthquake happening on the planet. The level of complexity and ambition is such that no one would believe a detection from just one observatory – so they built two. Hanford and Livingston are separated across the width of the US, such that any genuine signal arriving at one site should travel at the speed of light (unlike earthquakes and rain clouds) and arrive at the second site several milliseconds later. This subtle signal was detected for the first time in September 2015, days after a major LIGO upgrade, and it was detected at both observatories with exactly the right delay for the scientists to know instantly that this was a real detection.

The pattern revealed that gravitational waves had travelled to Earth from the collision of two black holes 1.3 billion light-years away.[5] Their collision sent ripples at the speed of light, eventually causing one of LIGO's arms to lengthen by just one-thousandth the diameter of a proton: a barely perceptible distortion with monumental implications.

The LIGO direct detection of gravitational waves was an astonishing accomplishment, one of the greatest landmarks in scientific history. However, it would not have been possible without the earlier observations of the Hulse–Taylor binary. While the waves from these neutron stars are too faint for LIGO to detect even today, their decaying orbit, meticulously tracked through pulsar timing, provided the first confirmation that energy could be carried away by gravitational waves. Nestled in the Puerto Rican jungle, the Arecibo Observatory achieved this with basic facilities at a fraction of the cost of modern detectors. Without that indirect evidence, I'm not convinced any scientific council would have funded the billion-dollar facilities used in the LIGO observation.

In the previous chapter, we explored a lone pulsar in the constellation of Vulpecula, the Fox. Here, we've turned our attention to one within a pair of neutron stars in Aquila, the Eagle. Today, thousands of pulsars populate our catalogues, and the sky pulses continuously with their radio signals. They are cosmic timekeepers marking the passage of time on a universal scale.

With twenty-first-century technology, we can now observe large numbers of pulsars simultaneously over long periods, tracking the subtle ebb and flow of gravitational waves. Gravitational waves have a wavelength just as electromagnetic radiation does, which is determined by the event that triggered their release. Earth-based detectors excel at measuring waves from events such as the collisions of stellar-mass black holes or neutron stars. However, they are blind to waves from larger phenomena, like the mergers of supermassive black holes, whose wavelengths far exceed the size of our planet. For such colossal ripples, the entire Earth stretches and compresses uniformly, rendering any distortion undetectable . . . from the ground.

As we know the exact positions of many pulsars thanks to their radio emission, we can use them as a colossal gravitational wave detector, known as a *pulsar timing array*. The steady ticking of a pulsar is as reliable as an atomic clock, enabling us to measure minuscule changes in distance in virtually any direction. Instead of mirrors at the ends of tunnels, we have pulsars scattered across the sky; instead of lasers and atomic clocks, we rely on the rhythmic pulses of radio light emitted by spinning neutron stars.

A pulsar timing array can detect much longer-wavelength

gravitational waves than Earth-based detectors because it spans vast distances across space. These immense baselines allow us to observe waves from supermassive black hole binaries spiralling around each other, which emit gravitational waves with wavelengths measured in light-years.

The mechanics of a pulsar timing array are strikingly similar to the Global Positioning System (GPS), a cornerstone of modern life and a staple of every smartphone. One might even call the pulsar array a 'Gravitational Wave Positioning System'.

When navigating unfamiliar terrain, all it takes is a tap on a map app to explore a new neighbourhood – or input a destination for precise directions. If we get lost, a constellation of satellites orbits above, ready to pinpoint our location and guide us onwards. The GPS network consists of a few dozen satellites, each equipped with a radio antenna and an atomic clock synchronised with an identical clock at an Earth-based ground station. These ground stations maintain constant radio communication with the satellites, measuring how long each transmission takes to travel between the ground and the satellite. Since the speed of light is well known, this data allows for precise calculations of each satellite's position.

When you enable location services, your phone receives time-stamped signals from satellites above the horizon. Your position is then calculated by comparing the arrival times from multiple satellites. This triangulation can pinpoint your position on the planet to within a few metres, easily placing you in the right room, let alone the right building. But GPS is not just used to find the nearest coffee shop within 5 m. It's also used to track potentially hostile air targets, prevent satellite collisions with spacecraft and guide soldiers through dangerous terrain with millimetre accuracy. For everyday tasks like finding coffee, an uncertainty of a few metres is tolerable; for avoiding landmines, precision can be a matter of life and death. The more accurate the timing of radio signals between ground stations, satellites and devices, the more detailed the map.

Pulsar timing arrays work on the same principle but on a vastly larger scale. Instead of satellites, astronomers use the most reliable pulsars as their reference points. They are often isolated, or with well understood binary companions, ensuring that any change in their pulse timing must result from a distortion in spacetime itself. The gravitational waves these arrays detect originate from the background hum of supermassive black hole binaries scattered across the Universe. These binaries create whirlpools in spacetime, sending long-wavelength ripples that merge, overlap and cancel out, producing a gentle, deep-sea roll rather than a tsunami.

If we were measuring a shift in just one pulsar, any quirk of our instrument, or anomaly in the pulsar timing, could overcome the measurement itself. But by measuring shifts in an array of tens of pulsars, we can average away chance errors and make a much more precise measurement. For example, a pulsar glitch might be a potential explanation for a change in timing in one pulsar, but it is a highly unlikely explanation if that change of timing is seen in ten pulsars, or even two.

In June 2023, a collaboration of radio astronomers called the North American Nanohertz Observatory for Gravitational Waves (NANOGrav) announced the first detection of this gravitational wave background, a 'cosmic hum' of spiralling supermassive black holes.[6] They combined observations of multiple sets of pulsars, taken from multiple radio telescopes around the globe. The pulsars were observed on a schedule, for example once every week for fifteen minutes, for upwards of *two decades*. This is an experiment that requires a long baseline, or 'control', data set. Years of observations of the same pulsars are required for comparison, when seeking to account for any tiny changes in pulse arrival times.

The result is a powerful piece of evidence for Einstein's prediction that gravitational waves should ripple continuously through the fabric of spacetime. Relatively nearby mergers of black holes send out gravitational waves strong enough to be detected individually by observatories such as LIGO. But as we have seen with the Hulse–Taylor binary system, general relativity doesn't just apply to black holes. It applies to the tiniest system of two stars slowly leaking gravitational energy as they spiral closer. These sources of gentler gravitational waves, and countless gigantic events that are simply so far away their waves have dampened considerably, blend together.

If you ever visit a radio telescope, it's likely to be observing a pulsar at some point. Many major radio telescopes regularly check in on a set of pulsars whenever they have spare time in their schedule. For astronomers, these spinning neutron stars are like old friends, their steady tick-tock familiar across entire careers.

The Hulse–Taylor binary in the Aquila constellation does not feature in the rosters of pulsar timing arrays because its shifting orbit obscures any subtle ripples in spacetime. Nevertheless, this pair of neutron stars paved the way for this ingenious use of pulsars to study gravitational waves. While Hulse and Taylor didn't directly detect the gravitational waves produced by the binary's spiralling motion, they used the pulsar's radio emissions to provide a blow-by-blow account of the system's evolution.

The cosmic bout cannot last for ever, though. With each tick of the clock, the neutron stars get closer. Just a few metres every year, a slow-motion collision, yet unstoppable. There will come a day, 300 million years in the future, when the two exhausted fighters will each throw their final knockout punches. Their death match ends not with a winner, but with a transformation. As the two stars merge, they will trigger a *kilonova*, at once one of the most destructive and creative processes in the Universe. Yes, much of the stars' mass will be lost for ever into the depths of a black hole, but further out, the violence and turbulence of that merger will instigate cosmic alchemy.

Kilonovae, the explosive outcomes of neutron star collisions, result in such high energies and temperatures that they can mimic and even go beyond the fusion that occurs within stars. During these brief explosions, some of the heaviest and rarest elements in the Universe are formed, elements such as gold, platinum and uranium. By studying the levels of these elements in Earth's crust, scientists can infer where and when such cosmic events occurred. Layers dated to 3.5 to 4.5 million years ago show unusual enrichments of certain heavy elements, most likely from a nearby supernova. But some recent models suggest that these concentrations are best explained by a long-ago kilonova event a few hundred light-years away bombarding us with heavy elements.[7]

A kilonova explosion is bright across the electromagnetic spectrum, including in the gamma rays. Should your planet stray too close, this radiation would incinerate the life that walks it. Even if one were lucky enough to avoid those death-rays, the high-energy particles from a kilonova a mere ten times closer, or thirty-five light-years away, would eradicate the ozone layer, leaving all life vulnerable to a lethal suntan for millennia: an extinction-level event. Thankfully, the Hulse–Taylor binary is far away enough that its eventual kilonova will pose no threat to Earth. But it's still close enough that, when the last punch is thrown and the neutron stars finally collide about 300 million years from now, the explosion will be bright enough to be visible across Earth's sky, and strong enough to produce measurable gravitational waves.

It will be the end of a long story that, until that point, was only available to radio astronomers. There was a gap of almost a hundred years between the publication of Einstein's theory of general relativity and the development of gravitational wave detectors such as LIGO. During that time, radio astronomers took up the challenge of testing those theories to their limits, and proving Einstein right, confirming that gravitational waves ripple through spacetime, just as he predicted.

And yet Earth, floating within this churning, rippling, roiling spacetime, goes on spinning. The Universe is so immense that, even though supermassive black holes are merging in every direction out in the wider Universe, we hear them just as a background noise of gravitational waves. Life goes on, at least in the Solar System, perhaps even in other star systems, and other galaxies.

11

EXOPLANETS

> Two possibilities exist: either we are alone in the Universe or we are not. Both are equally terrifying.
>
> Arthur C. Clarke

What is special about Earth? The Sun is one of several hundred billion stars in the Milky Way and decidedly average; our galaxy is just one standard spiral among countless others; our Universe might even be one within a multiverse of possibilities. Even within our Solar System, our planet is one of a brood of eight. In many ways, we are not special at all. Yet Earth holds a distinction: it is the only planet we know of that hosts life.

Speculation about life on the Moon, Mars and Venus flourished through the early twentieth century, only to be dispelled in the latter half as, one by one, these worlds revealed themselves as inhospitable. The *habitable zone* in a star system is defined as where liquid water can form on a planetary surface. Too close to the star and the temperature is too hot. Too far away, and it is too cold. This annulus of space around a star is also known as the *Goldilocks zone*, where the temperature is 'just right' for the eponymous fairytale character to eat her porridge. It won't surprise you to know that Earth sits squarely in the Solar System's habitable zone, which extends just shy of the orbits of Venus and Mars. Within Mercury's orbit, Goldilocks' porridge would evaporate, while beyond Mars, it would freeze.

The rule of thumb for astronomers seeking Earth-like planets used to be simple: location, location, location. Larger, hotter stars will have wider and more distant habitable zones, while smaller, colder stars will have narrower zones, closer to the star. But studying the Solar System planets, we have realised that Earth's habitability emerges from an extraordinary convergence of many more factors. Its location in the Solar System matters, yes, but we also need a stable atmosphere that traps oxygen and filters UV, X-ray and gamma wavelengths; tectonic activity that stabilises the climate by trapping and releasing carbon dioxide; and a magnetic field shielding us from the solar wind and cosmic rays. Venus and Mars are at the edges of the Goldilocks zone, but both are utterly hostile to life. Venus, with its dense, toxic atmosphere, would crush and melt even the sturdiest of porridge bowls within seconds. Mars suffers the opposite fate. Its thin atmosphere exposes its surface to meteorite impacts, temperature extremes and fatal radiation, hardly a relaxing place to eat one's breakfast, even if Goldilocks were wearing a spacesuit to help her breathe.

Our understanding of what makes a planet habitable is in constant flux and increasingly requires us to understand how planets are constructed, brick by brick, or rather pebble by pebble. Within the Taurus constellation, radio telescopes have captured images of the very earliest stages of planetary formation around baby stars. We now have photographs of those precious early years when a star system takes its first step into planetary formation and speaks its first words in the form of long-wavelength light. Once we understand the environment in which a habitable planet can form, we have a powerful filter with which to search the hundreds of billions of planetary systems surrounding us. So powerful that we may well find out whether Earth is truly special, truly alone, within our own lifetimes.

Our Solar System is not so thoroughly explored as to rule out more primitive, microbial life: either extinct in the red dust of Mars, or alive in oceans underneath the icy surface of Saturn's moon Enceladus. But, having flown missions to Pluto and beyond, astronomers can confidently conclude that Earth is the only planet able to host complex, intelligent life, at least within the Solar System. Our unique habitat made Earth appear special . . . but profoundly alone. Perhaps, some feared, the conditions for creating life were so unique that Earth's uniqueness extended beyond the Solar System, into a lifeless Universe. Perhaps even creating planets with stable orbits was near impossible, such that even uninhabitable planets like Venus and Mars were hard to find. But everything changed in the mid-1990s with the discovery of the first planets around other stars, *exoplanets*. This is astonishingly recent history. If you are over thirty years old, then you will have been born into a world with just nine planets (since Pluto was yet to be demoted). Readers under thirty, however, are unlikely to raise an eyebrow at the idea of exoplanets, having grown up in a world where we knew with absolute certainty that our Solar System was not one of a kind.

In 1995, scientists observed optical light from the star 51 Pegasi, fifty light-years away in the constellation Pegasus, and noticed minuscule shifts back and forth in its wavelength.[1] This indicated the gravitational influence of an orbiting planet gently moving the star and inducing a Doppler shift. This planet is too small to see directly but large enough to tug its parent star into a slight wobble, allowing astronomers to work out its metrics. This world, named 51 Pegasi b, is about half the mass of Jupiter but puffed up to about 1.3 times its radius. This was certainly no Earth-like planet, but its discovery transformed exoplanetary studies into a data-driven observational science. With advancing techniques and ballooning funding, astronomers soon uncovered a profusion of exoplanets: tens, then hundreds, then thousands.

Wherever astronomers look, they find planets. Today, we estimate that 70–90 per cent of all stars host at least one planet, and pretty much all Sun-like stars have their own planetary family. With several hundred billion stars in the Milky Way alone, the odds for there being more than one life-sustaining planet start to stack up. A habitable, life-sustaining planet may be rare, but even the rarest trading card appears multiple times when you purchase enough packs. Astronomers have several hundred billion packs, or star systems, to open. However intricate the requirements for hosting and sustaining life, the vast number of star systems out there suggests that life, somewhere, will emerge. Earth, remarkable as it is, may not be unique at all.

Against this backdrop, 51 Pegasi b holds a place in history as the spark that ignited the modern search for worlds beyond our Solar System. Its discovery rekindled humanity's hope of finding planets akin to our own. Places that might nurture life, just as we once dreamt Venus or Mars might. It was not the first exoplanet to be discovered, however, contrary to popular belief even amongst astronomers. That accomplishment belongs to the radio astronomers who detected an exoplanet called PSR B1257+12 B.

Three whole years before the discovery of 51 Pegasi b, in 1992, Aleksander Wolszczan and Dale Frail announced their detection of PSR B1257+12 B orbiting a pulsar rotating once every six milliseconds.[2] The planet was detected because of a slight Doppler shift in the expected radio signal of its parent neutron star, received at the Arecibo radio observatory.

Optical methods of exoplanet detection work by spotting subtle changes in a star's light, whether it is the wavelength shifting as the star moves under the planet's pull, or its brightness dimming as the planet passes in front. These methods are inherently biased toward finding planets that cause the most noticeable disturbances: massive, close-orbiting worlds such as 51 Pegasi b. Pulsar timing, by contrast, uses its unrivalled precision signals to detect planets of all sizes. PSR B1257+12 B has a mass similar to our Moon, just 2 per cent that of Earth.

The trouble, of course, is that planets detected via pulsar timing are, by definition, orbiting pulsars, which are among the most hostile stars imaginable. These planets are rocky worlds which will likely have been stripped of any atmosphere, if they ever had one. Aside from their bright radio beams, pulsars emit their light mostly in the shortest wavelengths, such that any inhabitants would experience lethal doses of X-rays and gamma rays, and at most only a hint of optical light. Life cannot thrive, or even survive, in such conditions. This harsh reality explains the scientific community's muted enthusiasm for radio astronomy as a method for detecting exoplanets; pulsar planets are dead, barren worlds and the herd curiosity remains firmly in finding Earth-like planets capable of supporting life.

The thousands of exoplanets discovered via both optical and radio means have demonstrated immense, and unexpected, variation. Among these are 'hot Jupiters', massive gas giants that are closer to their stars than Mercury is to our Sun, and 'super-Earths', rocky planets up to four times larger than our own. There are 'mini-Neptunes' with thick, gaseous atmospheres and 'super-puff' planets with densities as low as candy floss. 'Lava worlds' consist of oceans of molten rock covering the surface, and 'water worlds' are blanketed entirely by global oceans. There are planets where storms of glass shards whip around the surface, worlds where the surface is more likely to be made of diamonds than of rock, and even frozen, orphaned planets which seem to be floating freely in space, with no parent star at all. Each new exoplanetary discovery reshapes our understanding of how planets form and evolve, expanding the boundaries of where, and how, we might find other habitable worlds.[3]

But, to date, no planet beyond Earth has been definitively identified as habitable. It is not simply a case of looking for the third rock out from a Sun-like star. Our Solar System, with its relatively stable, well-spaced planets, is just one arrangement among countless cosmic possibilities, many of which we cannot explain with existing theories of planetary formation. Our attention is understandably drawn to finding the finished product of this formation process, a habitable planet, the culmination of a complex recipe involving rare ingredients. But to understand how planets form and what makes them habitable, we first need to study what goes on in the cosmic kitchen.

To find young planets, we need to look for young stars, and a molecular cloud is the perfect stellar nursery. These clouds are the densest regions of interstellar space, containing gases, icy grains and heavy elements expelled from supernovae. Mostly, though, they are dominated by cold gas in the form of molecular hydrogen, two hydrogen atoms bonded together. Within these clouds, the temperature is cool enough and density high enough that new generations of stars can begin to form from the remains of the previous generation.

The Taurus Molecular Cloud (TMC) is the closest large stellar nursery to Earth at 'only' 430 light-years away. This is still closer than many of the stars which make up the outline of Taurus, the Bull constellation. The exception is Aldebaran, the star marking the eye of the bull, which sits about sixty-five light-years from

Earth. You can identify Aldebaran quite easily because it is one of the brightest stars in the night sky, and it has a distinctly orange hue, much like Mars. This star is in the red giant phase, the end of a star's life. Despite only being marginally more massive than our Sun, its radius is forty-five times larger. Eventually its outer layers will dissipate, leaving its core as a white dwarf. Stars evolve along a prescribed sequence: from protostar to main sequence once stellar fusion is ignited and stable and then, once the fuel has run out, to stellar remnant. Aldebaran is currently transitioning into the final stage, while the stars of the TMC are just starting their journey.

The TMC is large, stretching approximately thirty light-years across Taurus and its neighbouring constellation, Auriga. All told, it contains about 30,000 solar masses of raw materials, ripe for stirring up into shiny new stars and potentially life-harbouring planets. Within this cloud is the archetype of the kind of star we are looking for: a T Tauri star. T Tauri stars appear in stellar nurseries all over the sky, but they are named after T Tauri, a specific star and the first observed of its kind, in the TMC. This class of star is less than 10 million years old, so young that they have not yet started to fuse hydrogen. Instead, their luminous surfaces result from the gravitational contraction of the cloud heating the gas such that it radiates energy in the form of light. Eventually, the core will heat to the high temperature required for fusion to ignite. Until then, the contracting outer layers of gas are met with violent stellar winds, and angry eruptions of radiation.

During the initial collapse stage, a protostar is heavily obscured by the cloud of infalling material. As the gas and dust rotate along with the rapidly spinning star, the material is flattened, becoming a *protoplanetary disc*. The star heats the grains of dust in the protoplanetary disc, which in turn give off thermal radiation at a wavelength related to the temperature and, crucially, the size of the grain. With radio telescopes, we are able to see the whole disc and also filter the system for different sizes of dust grains by tuning in to different wavelengths.

In 2014, the Atacama Large Millimeter/submillimeter Array (ALMA) revolutionised the study of protoplanetary discs, producing the first direct image of a planetary nursery.[4] ALMA consists of just over sixty portable 12-m-diameter radio dishes in the Atacama Desert of northern Chile. With the help of a purpose-built truck, the dishes can be rearranged to satisfy different scientific objectives that require a widespread interferometric array for high-resolution observations (dishes set far apart), or one small composite 'dish' to pick up the faintest, more diffuse signals (compact spacing).

The dishes were carefully constructed and then left high and dry up a mountain. While most radio telescopes can see straight through clouds, millimetre telescopes can struggle even on a relatively clear day, because shorter radio wavelengths get absorbed and scattered by the moisture in the atmosphere. So ALMA, which even observes sub-millimetre wavelengths, was installed at an altitude of 5,000 m in one of the driest places on Earth, leaving the least-possible amount of atmosphere between it and the stars.

Technically, the short wavelengths that ALMA observes at are part of the microwave section of the electromagnetic spectrum or, as the website puts it, 'cosmic light that straddles the boundary between radio and infrared'. But it uses antennas and dishes, as opposed to the optical instrumentation of lenses and cameras, so for the purposes of this book, I have included ALMA as an honorary associate of Team Radio. ALMA chose the TMC as the target of its first protoplanetary observations. Pointing its dishes towards the HL Tauri system, a T Tauri star, ALMA gathered the faint millimetre-wavelength thermal radiation emitted by the grains within the protoplanetary disc. The resulting image shows the system out to 200 astronomical units from the central star, equivalent to observing the Solar System out to the outer edge of the Kuiper belt, beyond Pluto, beyond even the *Voyager* spacecraft.

At first glance, the image looks like a planet surrounded by rings: there is a bright yellow central circle, and then concentric circles of varying thickness around it, in shades of red and orange. It takes a moment to adjust our sense of scale and realise we are seeing an entire star system. We can choose any colour scheme we like to translate from the invisible wavelengths observed by the telescope, but the fiery hues chosen are fitting for a place of genesis.

At ALMA wavelengths, we see the thermal radiation only from the millimetre- and sub-millimetre-sized dust grains, rather than observing any planets directly. The potential existence of alien worlds is inferred by the absence of dust, and thus the absence of radio light, along specific orbits. The dark annuli between the orange circles are regions swept clean of that dusty material, either by magnetic fields, or by protoplanets carving through their paths. Like the Bull constellation within which HL Tauri sits, these planets keep their heads down and charge through their orbits, accreting any grains that get in their way, stopping for nothing.

Much like rolling a snowball to build a snowman, this process accelerates in a runaway effect. The larger grains accumulate ever more mass, carving lanes through the protoplanetary disc and leaving precious little material for their smaller, struggling neighbours, who are destined to remain for ever as asteroids. Planet formation progresses from dust to pebble to planet, of this we are certain. But the details of this progression are somewhat of a mystery. It's not as simple as grains fusing in stellar heat; temperatures plummet well below the melting point of rock not far from the star. There are mysteries to be solved when it comes to the very earliest stages of planet formation, and observing the small dust grains in a young protoplanetary system is a crucial step in piecing together what is happening.

HL Tauri is remarkably young, even within a class of stars known for their youth. Its luminosity and temperature together suggest that it is less than 1 million years old. When this system began forming, *Homo erectus* was braving an ice age, alongside early mammoths and sabre-toothed cats. It seems a long time ago, but it is a blink of the eye in cosmological timescales. Earth, for example, was formed 4.5 billion years ago. Yet, despite HL Tauri's youth, observations appear to show clear channels in the disc of surrounding dust and gas, suggesting that young planets may be roaming the system already. This is a challenge to our understanding of planetary formation, and suggests planets can form much earlier than we thought, despite the turbulent and unstable environment.

I could stare at images of protoplanetary discs for hours.[5] I feel as if I am watching an artist place the final touches on their work. There is that same sense of privilege and wonder. Some of the systems imaged by ALMA have shown massive gaps, others still have razor-thin channels. There are asymmetric discs and discs which are so perfectly symmetric they look as though they've been drawn with a compass.

ALMA is not the first telescope to image a disc directly. The Hubble Space Telescope also has the capacity to view protoplanetary discs, but in nowhere near as fine a detail – there are no channels visible within the dust. Hubble's discs are bird's-eye views of the building site, from far above – the shape of the freshly finished house is clear from the outline of the roof, but we cannot see the windows to work out the number of rooms, and the building materials are too small to see.

ALMA's antennas, on the other hand, can be tuned to multiple wavelengths of light within the long wavelengths it is designed to detect. As mentioned earlier, the dust's emission wavelengths rely on a few factors, including the size of the grain. By viewing the same system at multiple wavelengths, we can view how different sizes of grain are distributed in the disc, giving us insight into where in the disc planets begin to form first.

I can't help but link the diversity we observe in planetary systems to a greater possibility of finding a planet just like ours. It is hard for me to believe that an Earth-like planet has only formed once, even though creating a stable, habitable planet is not trivial. It requires the right star, the right orbit, and the right planetary neighbours. Now we have the power to look at planetary building sites across the galaxy, in vastly varying environments and in different stages of completion, so we can begin to assess the rarity of the Solar System.

Our fascination with planetary formation is not just about understanding how bits of dust stick together. We also have a vested interest in working out what makes a planet habitable, and how we can tell if a planet is habitable or not from its location and star type. Studying protoplanetary discs from the earliest stages helps us understand the subtleties of the Goldilocks zone. Yet even if we master the art of defining this elusive region, a planet within it might still lack the spark of life. Radio telescopes, however, have proven invaluable not only in revealing the ingredients for planet formation, but also in helping us find the ingredients for life.

We may each have our own view of what we need to survive: a glass of gin, or a monthly manicure, perhaps. When it comes down to it though, there is one necessity which trumps all others: water. Scientists have modelled multiple ways of creating life and, however exotic the biochemical pathways, it is only possible with H_2O. Water is the defining feature of our blue planet: it covers 70 per cent of the surface, to depths so unfathomable that some say we know our Moon better than we do our own oceans. It is the defining feature of us: over half of our body is water and the processes behind life require water to such a degree that, without a constant resupply of fresh drinking water, our bodies will fail in just three days.

It is no wonder that the search for a habitable planet is also a search for water.

A lot of the radio emission we have talked about so far has been from hydrogen, the atom that makes up the 'H' in 'H_2O'. When an electron is excited up to a higher energy level due to a collision or heating, it descends and emits a photon of a specific wavelength: 21-cm radiation is a wavelength specific to a hydrogen atom. Molecules also produce radiation of specific wavelengths; and water, that key ingredient to life, produces a well-known signature in the radio wavelengths.

Among the stellar nurseries of the TMC, radio telescopes like ALMA have detected water vapour within the protoplanetary discs of young stars. In the inner region of HL Tauri, where rocky planets likely form, ALMA has captured the first-ever high-resolution images of water vapour inside the Goldilocks zone, estimating its volume as several times that of Earth's oceans.[6] Such discoveries suggest the beginnings of worlds rich with the potential for life.

There are three contending theories for the origin of life in our Solar System. It could have evolved on Earth, in isolation. Or, the spark of life could have occurred somewhere like Mars, creating a basic microbial life that transferred to Earth on the back of an asteroid. Lastly, the ingredients for life, such as water and amino acids, could have originated far out in the Solar System and been introduced into the nurturing incubator of Earth via a comet. To study planetary formation, we observe protoplanetary discs at different wavelengths and put together the different stages. To understand the formation of life we must do the same, and study where the ingredients for life gather. The detection of water vapour in protoplanetary discs is a huge leap forward. Life, though, requires far more than water.

The interstellar medium (ISM) seems barren to our eyes, like a stretch of empty space. Yet it is teeming with atoms and molecules

– remnants of ancient supernovae or products of cosmic chemistry. Stardust, rich in diverse elements, forms the building blocks of planets and life. However, the extreme heat within stars prevents these atoms, floating nearby the star, from bonding into molecules. For that, we need the cooler, more tranquil environment of the ISM, which acts as a sophisticated factory, crafting the complex molecules essential for life.

Simple interstellar molecules are believed to form on the icy surfaces of dust grains within molecular clouds, such as the TMC. As a protostar emerges, its heat causes these icy grains to sublimate, releasing the molecules into the interstellar medium as gas. Here, the molecules collide and react, creating larger and more complex compounds. Their presence is revealed in the light we observe from background stars. As light travels through the ISM, specific wavelengths are absorbed by different molecules, acting as a filter and producing a distinctive forest of absorption lines in the spectrum. Many of these absorption lines fall in the radio wavelengths, such that over 80 per cent of interstellar molecules were detected by radio facilities first.

Molecular clouds float throughout the Milky Way, their rich compositions open to anyone with a radio telescope. For example, NH_3 (ammonia), the pungent smell in hair bleach, was discovered in 1968 by the Green Bank Telescope, which also discovered H2O in 1969. The National Radio Astronomy Observatory 11-m dish discovered SO_2 (sulphur dioxide), the smell of burnt matches and volcanos, in 1975, while the Institute for Radio Astronomy in the Millimetre Range (IRAM) 30-m telescope found NaCl (salt) in 1987.[7]

So far, over 250 species of molecule have been detected in space, many of them variants on the same, vital atoms. Molecules such as OH (an oxygen atom bonded to a hydrogen atom), CO (a carbon atom bonded to an oxygen atom), H_2 (two hydrogen atoms bonded together) are commonplace. More complex compounds, including NH_2CH_2CN (aminoacetonitrile), NH_2CHO (formamide) and C_3PO (friend of R_2D_2), have also been detected. The first two are thought to be precursors to the amino acids that underpin life on Earth. The third, of course, is my attempt at a *Star Wars* joke.

The recipe for life isn't written in any cosmic cookbook, but the prevailing theory suggests it began with simpler forms of DNA, perhaps a single strand of ribonucleic acid (RNA) capable of self-replication. RNA uses amino acids to construct proteins, requiring complex sugars to form its structure. Significantly, these C_3 sugars have been found in meteorites, supporting the hypothesis that such molecules were forged in the interstellar medium or the outer Solar System and later transported to Earth via comets.

Ever more complex molecules are being unveiled from the dense forests of absorption lines, bringing us closer to detecting the coveted amino acids that form the foundation of life. In 2008, the Arecibo Observatory's 305-m dish discovered prebiotic molecules methenamine and hydrogen cyanide in another galaxy, no less. Together, these molecules could form glycine, the simplest amino acid, though glycine itself has not yet been confirmed in space. However, the search for prebiotic molecules was never the primary role of the Arecibo dish. More specialised instruments, such as ALMA, excel at detecting the faint absorption signatures of the most complex molecules. In 2019, ALMA identified glycolonitrile ($HOCH_2CN$), a complex precursor to RNA, within the interstellar medium.[8] Ingredient by ingredient, we are gathering everything we need to understand how we came to be.

With all these chemical formulae and unpronounceable molecule names, it's easy to lose sight of what we're observing: the most basic building blocks of life. It is profoundly wonderful to me that a radio telescope like ALMA can detect not only the raw materials of planets but also the ingredients for life that might walk upon them. The TMC hosts stellar nurseries and protoplanetary discs just millions of years old, if not younger. We can glimpse the origins of fully formed exoplanets, helping us understand where it is too hot, too cold, or just right. Beyond that, radio telescopes hint at whether the porridge of life has enough water, enough salt and enough sugars: the right recipe for life to thrive.

Earth, and planets like it, form step by step, from pebble to planet, and so does life, from sugars to amino acids to RNA. Even the simplest life needs ideal conditions in order to survive, replicate and thrive. The problem of how to create life seems insurmountable, a chance alignment of circumstance. Except that in just the last twenty-five years, the known number of exoplanets has jumped from 30 to well over 5,000. Everywhere we have looked in our nearby surroundings, we have found planets, such that across the Milky Way, there are likely over a hundred billion exoplanets.

The sheer number of exoplanets means that even the rarest conditions for life are likely to have been met more than once. Even in Taurus, aside from the young molecular cloud, there are multiple established star systems with potentially habitable planets. As we look up at these exoplanets, it's worth considering that somewhere out there, life may have already emerged, looked up to the stars, and wondered if they are alone. Perhaps they have already turned their radio telescopes towards the Solar System, having picked out a habitable-looking blue and green gem within a family of eight. Perhaps they have been sending messages of curiosity and hope, and are even now patiently waiting for a reply. Maybe it's time we checked our messages.

12

CONTACT

> It is possible that the future of human civilization depends on the receipt of interstellar messages.
>
> Carl Sagan, 'The Quest for Extraterrestrial Intelligence', May 1979

This chapter is about the search for extraterrestrial intelligence, or SETI, a perfectly respectable area of astronomy without any kind of controversy. Okay, okay, it is about aliens. Aliens of the kind we grew up with in cheap paperbacks and cult TV shows: green skin, extra arms, telepathy, X-ray vision. Logical Vulcans, furry Tribbles, moody Klingons . . . all are welcome. This is the chapter where we talk about the science fact behind all that science-fiction, and the hunt for intelligent life on planets other than Earth.

During public lectures, I'm often asked, 'Do you think aliens exist?' I am always a little taken aback, as is the audience when they hear my emphatic answer: 'Absolutely. I have no doubts at all.'

My surprise comes about because, as a radio astronomer, I have no doubt that life is out there, and I sometimes forget that there can even be a question about the matter. There is little controversy amongst astronomers, not since the avalanche of exoplanet observations showed us that it was only a matter of time before we found Earth 2.0. Even if habitable worlds are scarce or the emergence of life is rare, the sheer number of planets in the Milky Way, let alone the Universe, makes it implausible to think we are alone. I am not engaging here in a wilful misunderstanding of what we mean by 'alien'. When I say I believe there are aliens, I mean intelligent, complex lifeforms, not just basic, microbial life.

The audience is often silent for a beat after I have answered, not expecting such certainty. People expect heated debate over the question of alien life, I think, because there is a conflation, even in the scientific world, between a belief in alien life and a belief in UFOs, or alien abductions. 'Have aliens visited our planet?' is a fundamentally different question, and one which I can answer just as emphatically: 'Absolutely not. I have no doubts at all.' There is no evidence for interference by extraterrestrial life on Earth. No aliens building the pyramids, no UFOs creating intricate crop circles, and no secret government cover-ups. Many UFO sightings turn out to be Venus gleaming on the horizon, optical illusions or simply shaky camera work. One only need note how UFO sightings have not increased in line with smartphone ownership, and the ability to capture supposed otherworldly events. These claims belong firmly in the realm of conspiracy theories.

SETI, however, stands apart as a rigorous, science-based pursuit, almost exclusively led by radio astronomers. For SETI researchers, the question isn't whether life exists elsewhere – we take that as a given. We want to make *contact*. We want to find out where these aliens live and communicate with them across the light-years, using radio telescopes as our telephone.

Radio astronomy has its origin in communication: remember Karl Jansky discovering the first 'extraterrestrial' radio waves, from the black hole at the centre of the Milky Way? This only happened because those radio waves were interfering with the clarity of transatlantic phone calls. It is therefore quite fitting that radio telescopes born from the need for humans to communicate amongst themselves are the most likely tool by which we will make contact with an intelligent extraterrestrial species.

Today's radio telescopes would be astonishingly sensitive to the whispers of intelligent life. Instruments like the 100-m Green Bank Telescope in West Virginia are capable of detecting transmissions from alien planets up to 400 light-years away, assuming they use a transmitter comparable to Earth's now-defunct Arecibo Observatory. A transmitter a thousandfold more powerful would extend that range to thousands of light-years, potentially reaching billions of stars.

Yet there has still been no contact. Disappointing, perhaps, but taken alone, this silence does not mean that extraterrestrial intelligence is uncommon. It could be that there is no one out there, but it could also just be difficult to make contact. Sending a powerful message into the galaxy takes energy, and any species would have to choose their target – we might not be it. They are also likely be confined to only sending their message for a short time-span, whether because of political cycles, energy resources or telescope timetabling. Because our limited telescope time also restricts where and when we can search, we might simply turn our focus to a planet just as their radio astronomers have gone on annual leave. So where do we dare to look, knowing that, just out of sight, we might miss a message that will never be repeated?

There are promising planets across the sky, but I want to point us initially in the direction of Aquarius, the constellation of the 'water-bearer'. On top of the outline of stars, most artistic renderings depict a man pouring water from a vessel onto the ground. This beautiful prince served ambrosia and nectar to the Greek gods before being immortalised in the stars by Zeus as a reward for his service. In Roman cultures, Aquarius symbolised life-sustaining rain. Its appearance in the sky marked the annual flooding of the Nile, Tigris and Euphrates rivers, so the Egyptians and Mesopotamians associated it with fertility and renewal.

Amongst the stars representing the flow of water from Aquarius' cup, there is a small red dwarf star, originally designated 2MASS J23062928-0502285. It is 100,000 times fainter than any star we can see by eye, but visible to our large optical observatories, including the Transiting Planets and Planetesimals Small Telescope (TRAPPIST) at La Silla Observatory in Chile. This telescope identifies exoplanets by surveying stars and looking for tiny, regular dips in their brightness, corresponding to a planet passing, or transiting, across the surface as it orbits that star.

In 2016, TRAPPIST discovered that the light from 2MASS J23062928-0502285 suggested the presence of three planets. With the help of other telescopes, seven planets were eventually confirmed.[1] Remarkably, not one but three of these planets appear to be within the star's Goldilocks zone, and therefore capable of hosting liquid water. To have found three potentially habitable planets in one system is cause for great excitement in itself, but in addition, the star is older than our Sun. This means that, aside from the primordial microbial life, there has been plenty of time for complex and intelligent life to evolve too.

The star 2MASS J23062928-0502285, renamed TRAPPIST–1 in recognition of its newfound significance (and presumably to ensure more media-friendly press releases), hosts one of the most intriguing planetary systems in our cosmic neighbourhood. With potentially three habitable worlds, TRAPPIST–1 stands out as a beacon of hope in the search for extraterrestrial life. However, this system is far from a mirror of our own Solar System. To grasp its uniqueness, let us imagine life unfolding on one of its planets. Specifically, the fourth one out from its faint, red star.

Life on TRAPPIST−1e is cold and crowded. At first glance, you might expect there to be plenty of room. After all, it's a rocky planet the size of Earth and we have billions of humans aboard, let alone other species. But unlike our home, where life sprawls across almost the whole surface, existence on TRAPPIST−1e is confined to the 'twilight zone', a slender strip of habitable land the width of North America, encircling the planet from pole to pole.

Why the restriction? TRAPPIST−1e, like all its sibling planets, is close enough to its star that it is highly likely to be tidally locked, such that one rotation equals one orbit (146 hours, or six Earth-days). This means one side of the planet basks in perpetual daylight, while the other is cloaked in eternal darkness: just as our own Moon only ever shows one face to us. With little axial tilt, there are no seasons to break the monotony. The day or night is infinitely long, depending on which side you are on, and each year blends seamlessly into the next.

There is a 'twilight zone' straddling the terminator, the delicate line between light and shadow. Here, residents endure semi-darkness and fierce winds driven by the planet's starkly bipolar weather system. The most coveted real estate lies closest to the terminator, owned by the wealthy and powerful. Those less fortunate live nearer to the edges, where life struggles against either dehydration in the scorching light or freezing in the frigid dark.

Even the sky feels crowded, though its beauty is undeniable. TRAPPIST−1e exists within a Solar System in miniature. All seven planets orbit within just 5 million km of their star, well inside what would be Mercury's orbit in our Solar System. From the twilight zone, sister planets loom larger than Earth's Moon, while their 'sun' takes up four times as much area on the sky as our own star.

TRAPPIST–1, a dim red dwarf, is the smallest type of star capable of nuclear fusion, barely larger than Jupiter and with less than a tenth of our Sun's mass. Its light is weak and ruddy, lacking the energetic blues and violets that give our sky its bright colour. On these planets, the sun never rises or sets: it hangs motionless in the sky, bathing the day side in a muted yellow-green glow. Though the star's light is intrinsically feeble, just a few hundredths of a per cent as luminous as our Sun, the planets orbit so close that TRAPPIST–1e, for example, still receives just over half the warmth Earth enjoys. Crucially, in the twilight zones, the temperature is enough to keep water liquid.

TRAPPIST–1e is the first planet in the system to be securely in the Goldilocks zone, but in the reverie we're entertaining here, it is not alone in hosting life. TRAPPIST–1f and TRAPPIST–1g, with their own habitable twilight zones, are teeming with various species. Travel between the planets is remarkably easy. Space exploration developed quickly under the pressure of overcrowding, and interplanetary travel takes just a few Earth-days.

The need for communication between these planets has necessitated a well-funded radio telescope programme, and radio astronomers are very highly regarded. Radio waves bounce happily between the planets, gliding through their nurturing atmospheres with news or gossip. Massive radio telescopes dot the twilight zone, their primary role to connect the planets, though some may also scan the stars for signals from other systems. After all, for the inhabitants of TRAPPIST–1, the concept of life on multiple planets is a given – three of their own worlds host life. Perhaps they've already identified Sol–3, the third planet of the distant Sun, as a candidate for habitability. Maybe, even now, they are sending us a message of friendship.

The details of their society are, of course, my imagination. The planets are real, their orbits are real, and their potential habitability is grounded in observation. Life may have evolved on TRAPPIST−1e, or it may not. But if it has, radio telescopes would almost certainly be their tool of choice for calling across the space that separates us.

Since the 1960s, SETI scientists have possessed the technology to detect radio signals from potential alien civilisations. Yet their efforts have been stymied, not by technological limitations, but by an enduring stigma. Despite, or perhaps because of, the widespread fascination with the prospect of alien life, SETI has often been regarded as a fringe pursuit within the scientific community. This perception has made securing funding a perennial challenge, with telescope time equally scarce. As a result, SETI researchers focus their limited resources on targeting the most promising regions of space where intelligent life might exist.

The TRAPPIST−1 system is significant because it meets several criteria necessary for intelligent, contactable life. These criteria act as a filtering system, narrowing down the billions of planets in the Milky Way to a manageable selection of potential civilisations whose electromagnetic emissions could be detectable. Imagine the Universe as a massive online store. To make sense of it, we need filters to refine the search. First, we consider the rate of star formation, excluding volatile objects like pulsars and black holes, which cannot host stable planetary systems. Stars like our Sun pass this filter, they are long-lived enough for life to evolve, but they're relatively rare. Red dwarfs, such as TRAPPIST−1, dominate our sample because they are abundant and extraordinarily long-lived, making them excellent candidates.

From there, we assess the fraction of these stars that have planetary systems. For much of human history, we knew of only the five planets visible to the naked eye: Mercury, Venus, Mars,

Jupiter and Saturn. Uranus followed in 1781, Neptune in 1846, and Pluto in 1930, the last of which was discovered to be one of so many similar objects that it lasted less than a hundred years before its demotion to dwarf planet.[2] By the mid-1990s, with the discovery of exoplanets, we realised that planetary systems are common. Today, we know of well over 5,000 exoplanets, making this filter almost redundant, like searching a website named Skirts.com for 'skirts'.

Next comes the real challenge: how many of those planets are likely to have environments suitable for life? While many planets are too extreme – too hot, cold, irradiated or dry – initial estimates suggest that we might expect as many as one Earth-like planet for every three planet-hosting stars.

But what fraction of these habitable planets actually hosts life? This is far harder to estimate, as Earth remains our only data point. However, there's little reason to think Earth is special, so this filter likely only narrows the search, as opposed to ending it.

The next hurdle is intelligent life. This is a much harsher filter, as the emergence of intelligent life requires stable conditions over vast timescales. The 'rare Earth' hypothesis argues that Earth-like conditions are so improbable that intelligent life elsewhere might be vanishingly rare.[3] Others contend this view is too Earth-centric, and that intelligent life could evolve under different conditions.[4] SETI scientists believe that if intelligent life emerged here, it must have done so elsewhere given the sheer number of planets. Again, a narrowing filter, but not a deal-breaker.

If one is interested only in whether intelligent life is out there, this is where the filtering process stops, with a sizeable number of planets in the sample. But for us to *know* if there is intelligent life, we need to apply two more filters. For contact, we need additional criteria: detectability. Intelligent beings must produce electromagnetic signals strong enough to be noticed, whether intentionally or accidentally. This requires advanced technology, such as radio astronomy. While it's possible civilisations might avoid signalling out of fear, misunderstanding or disinterest, SETI scientists often see this filter as a minor obstacle. The assumption is that once intelligence develops, it will eventually produce detectable signals, even if all that's being broadcast is their version of a popular television programme.

Finally, we confront the most disheartening filter: longevity. How long does a civilisation remain capable of producing detectable signals? This parameter forces us to consider whether an intelligent civilisation can survive long enough for it to overlap in time with a civilisation that can detect their signals. Consideration of this filter brings with it a sobering reflection on humanity's own fragility. Climate change, nuclear threats, pandemics and other existential risks . . . we only have one example of a planetary civilisation, and we could be doing better. Perhaps other civilisations have been equally afflicted, reducing the chances of our societies overlapping.

This last filter is where optimism and despair collide. As we watch conflicts escalate and our planet's resources dwindle, we can't help but feel a twinge of shame in showing extraterrestrials how we have plundered this beautiful world. But SETI's work is underpinned by the hope that contact with another civilisation might unify humanity, offering a profound perspective on our place in the cosmos and providing the will to turn ourselves around as a species. Despite the challenges, even the slimmest chance of connection makes the search worthwhile, especially when the stakes are so high.

The above set of filters is collectively known in astronomy as the Drake equation, named after Frank Drake. In 1960, Drake attracted significant attention within astronomy and amongst the general public when he launched the first SETI effort, *Project Ozma*, named after the ruler of Oz in L. Frank Baum's novels.[5] Drake spent over a hundred lonely hours listening in the control room of a radio telescope in the National Radio Astronomy Observatory, having pointed it at Tau Ceti and Epsilon Eridani, two nearby Sun-like stars. He did not make contact, but his effort and enthusiasm were enough to get him noticed by members of the Space Science Board in the prestigious National Academy of Sciences. They wanted to convene a meeting to further research on extraterrestrial intelligence and called Drake to ask him who he thought would be interested: 'all of ten people,' as he later quipped.[6]

To keep discussions on track, Drake began organising the key questions that needed answering at the meeting, to tackle the subject scientifically. How many stars like the Sun are there? How many of those stars host planets? And so on. As he jotted them down, a pattern emerged: each question represented a filter, or a link in a chain, that must hold intact for intelligent, detectable extraterrestrial life to exist.

Drake assigned a symbol to each topic, eventually writing them out on a blackboard as a formula during the conference's opening session. This quasi-equation, he explained, was just a guide to assessing whether a search would be likely to meet with success. Jill Tarter, a pioneering SETI scientist, later described it as 'a wonderful way to organise our ignorance'. It was a roadmap, offering structure to one of science's most profound and nebulous questions: *are we alone?* Over the years, the equation has become iconic in SETI culture, adorning t-shirts, mugs and even tattoos (I've personally seen two).

The meeting itself, though small, was a resounding success. Attendees included Carl Sagan, the celebrated astronomer who later wrote *Contact*, and Barney Oliver, soon to be the first director of research at Hewlett-Packard. The neuroscientist John C. Lilly was invited after Drake jokingly noted that his list lacked 'someone who's actually spoken to an extraterrestrial'. Lilly captivated the group with his vision: cracking the code of dolphin communication might provide the key to understanding extraterrestrial languages. So inspired were the attendees that they temporarily named themselves the Order of the Dolphin.*

SETI was multidisciplinary in its assessment of whether life existed, but the search itself was firmly rooted in radio astronomy from its inception. The first conference was held at the Green Bank Telescope in West Virginia, one of the world's most iconic radio observatories, with half the attendees hailing from radio astronomy.

The Green Bank meeting provided SETI with legitimacy, momentum and a plan. As they dissected the terms of the Drake equation, the participants balanced optimism with realism. SETI, they agreed, was a long-term project. The search might take a century. But as they pointed out, with the technology at hand, it might just as easily succeed tomorrow. Now, sixty years on, with even more sensitive telescopes and computational power the Order of the Dolphin could only have dreamt of, the 'tomorrow' when we make first contact is closer than ever.

Nestled among the mountains of northern California, close to the tiny town of Hat Creek, lies the only radio observatory in the world built specifically for SETI. The road leading to the observatory is a dirt track, and a sign at its entrance politely but firmly requests visitors to turn off their mobile phones. After all, the scientists here are hoping for a call from E.T., not your mum. The observatory's remote location, a four-hour drive north of San Francisco, is no coincidence. Hat Creek itself is home to only about 250 residents, while the surrounding foothills shield the site from much of the radio interference of larger towns and cities. The wilderness offers an almost eerie silence, broken only by the faint hum of the telescope machinery.

* This moniker was quietly abandoned after it came to light that Lilly was taking LSD while swimming with dolphins in an attempt to talk to them.

The Allen Telescope Array (ATA) antennas might seem unassuming at first glance. Each is a small, 6-m dish with an odd, squashed canvas cylinder attached to one edge. If you unzip this section and peek inside, you'll find a small Christmas tree of metal that looks like a raygun from a sci-fi film. This is the antenna that captures the radio waves gathered and concentrated by the dish and feeds them to the cryogenically cooled amplifiers behind it. The design is known as an 'Antonio feed', after Franklin Antonio, a communications entrepreneur who donated $3.5 million to upgrade the array. This telescope has not been funded in the traditional way, by research councils, but by private individuals, notably the co-founder of Microsoft, Paul Allen.

Tilted skyward, forty-two dishes stand silently, listening to the stars as one. These antennas can rotate and tilt in unison, homing in on signals from a single star system with precision. But their graceful 'dance' can be dangerous to anyone wandering too close. Signs warn humans to keep a safe distance, and even the local livestock is accounted for: the dishes are designed not to tilt closer than eighteen degrees to their supporting poles, preventing curious cows from getting their heads stuck.

A key strategy at the ATA is to repeatedly survey a catalogue of a few hundred nearby stars considered likely to host habitable worlds.[7] The dishes align to observe a star system, listen intently for five to ten minutes, and then move to the next target. After cycling through the list of stars, the survey starts again, running day and night.

Why repeat the same observations? Because if an extraterrestrial intelligence is broadcasting, it's unlikely they're directing their signal solely at us. Transmitting signals across the vast distances of space requires a lot of energy. To maximise reach and clarity, it is best for an antenna to send its signal in a narrow, focused beam. If your dog has bolted out of sight, you wouldn't call out endlessly in one direction; you'd turn in all directions, shouting

'Bertie!' and using your hands to focus your voice, maximising the chances of your voice reaching them. In this analogy, we're the dog. Our alien species is unlikely to have the energy to power a telescope that can broadcast all the time and everywhere at once. Equally, we don't have the power to listen in every direction. That's why the telescope repeatedly cycles through its survey. The Allen Telescope Array is an attentive spaniel turning its head, ears pricked, to catch the faint cries of a cosmic companion.

Not only do we have to wonder which direction extraterrestrial signals might come from, but we must also consider what part of the vast electromagnetic spectrum they might be using to communicate. This is like being handed a radio with no preset stations or guidance about the frequency of transmission. As you twist the dial, scanning through static and chatter, you could easily miss a vital signal occurring temporarily on another channel.

The first SETI experiments of the 1960s could only observe one frequency at one time, due to the limitations of the technology. A lot of thought went into which frequency to choose. It had to be a frequency that the aliens would consider special in some way, such that anyone listening would have a good chance of guessing it. They settled on the frequency of 1,420 MHz, equivalent to a wavelength of 21 cm, the wavelength of radio light emitted by hydrogen atoms. This is a wavelength we have encountered before, when we explored how radio astronomers used it to map the spiral arms in the far reaches of the Milky Way. Hydrogen is by far the most common element in the Universe, making it a ubiquitous radio frequency in our Universe. So, if our hypothetical aliens are scanning the skies, they are bound to be tuned in to 1,420 MHz, just like us. It's like a government emergency broadcast going out across the frequency usually used by BBC radio stations – we choose it because we know they'll be listening anyway.

Thanks to advances in computing power and data storage,

modern SETI projects can now scan billions of frequencies simultaneously. While the 21-cm line remains a prime candidate, the ATA simultaneously searches frequencies of 500–11,000 MHz, drastically accelerating the search. No longer limited to guessing alien methodologies, SETI has transformed from a hopeful gamble into a broader, more efficient exploration of the cosmos.

Listening to this many radio channels at once generates a staggering amount of data, equivalent to downloading dozens of HD movies every second. This is far more information than can be stored on-site, and so hard decisions have to be made. Each observation of a stellar system is fed through an algorithm to decide whether to keep the data or delete it, based on its quality and likelihood of containing a signal. At this stage, we aren't even attempting to decode any incoming message. Instead, we're simply searching for the hallmarks of an alien signal. To put ourselves in the shoes (or paws) of the lost dog again, it doesn't matter if our owner is blowing a dog whistle to the rhythm of 'Space Oddity'. All we need to determine their location is the fact that they're blowing the whistle at all.

What the SETI algorithms look for is a signal slowly drifting through the frequencies. The motion of the star system relative to us introduces a Doppler shift in the signal. We can work out what that drift rate is for the target planetary system and look for exactly that pattern.

When an intriguing signal is detected, the data is saved, and a notification is sent to the scientist in the control room. Each one must be assessed to rule out other sources of drifting signals, such as lunar space probes. There's a strict protocol for what happens if an incoming signal passes the next set of tests: senior SETI scientists are automatically notified by text. They will descend on the ATA to perform their own checks, ensuring, for example, that the signal vanishes when the star system sets below the horizon. If the signal remains credible, the next step is to call other radio observatories that can also view the star system, begging for 'discretionary time', time reserved for urgent follow-up observations. If another telescope confirms the signal, well, I imagine all hell will break loose. Until that point, I can vouch for the fact that life at the Allen Telescope Array is very peaceful. The scientists are left watching the dishes watch the stars, and filling the silence with card games, while they wait, and wait.

Patience and luck are not qualities scientists typically like to factor into their experiments. Yet in this field, they are essential, shaping a certain kind of researcher. Almost every dedicated SETI researcher I have met carries a certain wariness. They listen intently, letting you fill the silences, as if gauging where your allegiances lie, or perhaps simply taking care not to come across as overzealous alien-chasers. When they do open up, though, they talk about the question of whether there is life elsewhere as the most profound inquiry of all. They are thick-skinned enough to weather scepticism, and content to bide their time.

Funding agencies, entrusted with public money, however, have often baulked at supporting a scientific endeavour with such a slim chance of success, especially within the tight constraints of a three- or five-year research grant cycle, no matter how 'profound' the potential discovery might be. Popular culture has worked against SETI, overall, with most people cordoning extraterrestrial life firmly on the 'fi' side of sci-fi. In 1978, Senator William Proxmire awarded SETI his infamous 'Golden Fleece' award, given annually to projects he deemed the most egregious waste of taxpayer money. The move was met with sharp wit from SETI figurehead Frank Drake, who publicly nominated Proxmire for honorary membership in the Flat Earth Society. The humour didn't save the programme, however, as Congress slashed SETI's funding entirely in 1982.

At this point Carl Sagan stepped in. Ever the master of persuasion, Sagan met with Proxmire in his office and argued that discovering an alien civilisation could provide humanity with a much-needed distraction from cold war fears of nuclear annihilation. Even more importantly, he suggested, evidence of another society surviving against the odds might inspire hope that humanity could avoid its own self-destruction. Proxmire was won over, and in 1983, SETI's funding was restored.

But this reprieve didn't last long. Just ten years later, in 1993, government funding for SETI was slashed completely, amounting to a saving of just eight cents per taxpayer per year. The Drake equation had a term added, a filter for whether the planet has an idiotic politician in charge. If so, the chances of contact became zero. It could have ended there, with the shutdown of radio telescope programmes snuffing out our chances of contact – like a phone running out of battery just as the call comes through. But SETI had a new champion, a pioneer in the field, who had already begun to walk a bold new path.

Jill Tarter began her journey in science studying engineering physics at college, eventually earning her PhD in astronomy from the University of California, Berkeley, in 1975. Her research focused on brown dwarfs, but the subject left her uninspired. Then she stumbled across the literature on SETI and was instantly hooked. Here was a subject that mattered, the exploration of which had the potential to change human history.

One of Tarter's early projects involved engineering a 'parasitic' device to attach to radio telescopes, harvesting data from other astronomers' observations so SETI scientists could analyse it for alien signals.[8] It was the scientific equivalent of asking for a friend's Netflix password. With SETI applications for telescope time frequently denied, Tarter and her colleagues needed creative ways to gather data. Her piggybacking technology offered a solution, but first, it had to be tested. Tarter and her team packed their equipment into reinforced metal suitcases, 'Suitcase SETI', and boarded long-haul flights to Canberra Deep Space Communication

Complex, host of several radio telescopes used to track NASA spacecraft. Brimming with excitement, they unpacked their pride and joy, plugged it in, and . . . the electronics blew. They had been so preoccupied with figuring out how to communicate with extraterrestrial intelligences that they forgot to plug their equipment into a travel adapter.

After the smoke cleared and the electronics were replaced, the system worked, proving the concept. Suitcase SETI was soon upgraded to 'Trailer SETI', a mobile home for their equipment that could be set up at any willing radio telescope. Willing telescopes, however, were scarce. The director of the Parkes Observatory in Australia offered Tarter six months of dish time for a $250,000 fee, and even then, the trailer had to travel by sea, a journey that took months. The search was taking too long and costing too much.

By this time, government funding for SETI had been cut, and Tarter had grown weary of hauling equipment from one telescope to another, dependent on the unreliable support of research agencies. She had already co-founded a new non-profit research organisation in 1984, the SETI Institute, in an effort to break away from their dependence on government funding. Now, a new idea was crystallising: it was time for SETI to have its own telescope. And who better to fund this visionary project than the curious millionaires and billionaires of Silicon Valley?

So, she spent the next decades tracking down donors, marching into the office of every rich person she could think of that might have an interest in the technology, the philosophy or the science of SETI. She spoke about the importance of the question *are we alone?* and echoed Sagan's vision of how SETI could unite humanity.

Her persistence eventually paid off, and the ATA became a safe haven for SETI researchers in 2007. No longer did they have to beg for telescope time or risk their projects being derailed by outside priorities. Though the ATA currently consists of just a few dozen dishes, far short of the 350 originally envisaged, the array is as expandable as a Lego project. You can just keep connecting new dishes when you have saved enough to buy the extra pieces. Even if the ATA never grows beyond its present state, it has already proven its worth. As we go about our daily lives, those forty-two dishes silently nod and turn in unison, scanning for signals from planets within fifty light-years, a region encompassing 5,000 to 10,000 star systems.

The greatest challenge with a SETI experiment is knowing when to stop. What constitutes a meaningful null result, how will we ever know we are definitely alone? Tarter uses the metaphor of filling a glass from the ocean. We do not assume there are no fish, just because the glass only contains water. I work on SETI with undergraduate students as a sideline to my research interests, and I'm happy to help fill a glass from the ocean now and again. But I couldn't spend all my time searching, I don't have the patience or resilience.

According to her biography, Tarter is haunted by the possibility that she might never know how close she has come.[9] What if they tried to call while she was on another line? What if she switched star systems just ten seconds before an extraterrestrial intelligence reached out? What if just one more observation was all it would have taken? It reminds me of always playing the same numbers on the lottery: it becomes almost impossible to skip a week, for a fear those exact numbers will come up. At a conference celebrating the remarkable success of exoplanet detections, Tarter allowed a rare moment of vulnerability to show, whispering, 'We didn't find anything.' For all the times she invoked the metaphor of filling a glass from the ocean, one suspects she held a deeper, quieter hope that she might beat the odds: that contact might happen in her career, or at least in her lifetime.

Perhaps it still will. It could be tomorrow. Knowing that, no

SETI scientist can ever truly walk away, and so the search continues.

SETI has endured largely as the plaything of wealthy private investors. For people who already have everything, SETI is the only lottery ticket with a meaningful prize: to go down in history as the person whose name is attached to the telescope that makes first contact.

In 2015, Soviet-born Israeli billionaire entrepreneur and physicist Yuri Milner announced the Breakthrough Listen initiative, a $100 million effort to reinvigorate the search for extraterrestrial intelligence.[10] The initiative gained the backing of prominent scientists, including Stephen Hawking. Breakthrough's strategy shifted away from building telescopes towards developing computational power and proprietary software that would allow the team to hop on and off many of the world's most advanced radio telescopes, scanning the sky faster, further, wider and across more frequencies. This method is referred to as 'Boutique SETI', in contrast to the rather more down-to-earth 'Suitcase SETI' and 'Trailer SETI' monikers.

Financial stability has eased the existential threat to SETI. Bolstered by Milner's substantial backing, the project is increasingly regarded as a credible investment. Crucially, donors are beginning to understand what it will take to make contact a reasonable prospect within their lifetimes: seven-figure contributions. One can only hope that Breakthrough does not auction off the rights to the first message: 'We come in peace, thanks to the refreshing effect of Coca-Cola.'

In 1974, Frank Drake was the director of the Arecibo Observatory. Like Tarter, Drake lived and breathed SETI, and he had become impatient at the pace of the search. SETI is passive. It is about listening and intercepting signals that extraterrestrial life has sent, purposefully or not. Drake wanted to take the initiative, to beam a message out ourselves. And so, METI (messaging extraterrestrial intelligence) was born.

Two years earlier, the *Pioneer 10* space probe had launched, carrying a gold-plated plaque with etched diagrams and info-graphics about Earth and humanity. Drake decided to send his signal using the 305-m dish's antenna as the transmitter, at a celebration honouring the recent renovation of the dish.[11] *Pioneer 10* would leave the Solar System in 1983, eleven years after its launch. A message coded in radio light, meanwhile, would travel at 300 million m/s, overtaking *Pioneer 10* in only an hour. The message would be out of the Solar System in less than a day.

The message consisted of 1,679 binary digits, each representing one of two possible values. Frank Drake envisaged encoding a picture in this way, much like a 'paint by numbers' canvas. A radio tone at one wavelength would indicate a solid square, while a tone at a second wavelength would represent an empty square. Since the message would be transmitted as a long sequence of radio bursts, the challenge for any receiving extraterrestrial intelligence would be to determine how to arrange this list into a meaningful 2D image. The image contains a wealth of information: heavily pixellated diagrams of a human and the Arecibo dish, for example, along with a representation of our Solar System, and the major chemical elements making up our DNA.

Drake selected the number 1,679 deliberately because it is a prime number and also the product of two other primes: 23 and 73. Prime numbers are significant within mathematics, and Drake hoped that any intelligent beings receiving the flow of data would experiment with reshaping the sequence in configurations involving these primes. If they did so, they would discover the intended image, unlocking the deeper meaning of the message.

Drake also converted the message into audio, allowing the audience gathered next to the dish in the Puerto Rican jungle to listen in real-time as the two-toned signal was broadcast into space. For three minutes, Earth transmitted a signal 10 million times brighter than our radio Sun, making us unmistakably visible to any extraterrestrial intelligence with their own Arecibo within the target area.

I've listened to the signal myself. The *Moonlight Sonata* it certainly is not, but if you close your eyes and reflect on the poignancy of us announcing our presence to the cosmos, it's deeply moving, and I can only imagine it brought a tear to the eye of more than one person in the crowd.

The message was aimed at M13, a dense cluster of stars (and presumably planets) about 21,000 light-years away, in the Hercules constellation. This system was chosen partly for the press-release optics of its dense population of potential life-bearing star systems, but mostly because it was above the dish at the time of the ceremony. Awkwardly, by the time the Drake signal reaches the intended coordinates, M13 may have drifted out of the way of the tight beam of the signal. The act was symbolic but still, the first intended message is out there, and we cannot undo it now. By the time the 1974 ceremony lunch was over, that signal had reached the orbit of Pluto, and by now it is over fifty light-years away, about as far from us as 51 Pegasi b, the first exoplanet to be discovered around a Sun-like star.

Drake did this all without needing anyone's permission, and many saw his act as setting a dangerous precedent. Why should the decision to contact alien life be autocratic when any risk is shared by everyone on the planet? England's Astronomer Royal Sir Martin Ryle reacted with scathing letters to the International Astronomical Union demanding a ban on any future messages, since 'for all we know, any creatures out there might be malevolent – or hungry'.[12]

Drake and, by this time, the majority of SETI scientists considered the argument moot. Electromagnetic radiation has been escaping Earth's atmosphere since radio and television were invented. Rather charmingly, the distance these emissions have travelled is called the 'I Love Lucy' radius, after the hit 1950s sitcom. The wide popularity of the series required it to be broadcast repeatedly and over a wide area, causing significant emissions into space. It has been over seventy years since the first episode aired, and so if there are aliens with the technology to pick up what would be a very weak signal, weaker than the leftover radiation from the Big Bang, they could be watching that programme . . . as long as they are less than seventy light-years away. Maybe, even now, someone on a planet in the TRAPPIST–1 system is listening. What might they hear from us? The sounds and images of 1985, perhaps: a curious mix of synthesisers, *The Goonies* and *Back to the Future*.

Whether we're at risk of invasion depends on two things: first, whether they manage to develop interstellar travel, and second, whether they decide such an endeavour is worth the colossal energy cost. Imagine our own situation – would humanity, in the midst of an energy crisis, unite to devote gargantuan resources into building and launching an interstellar colonisation fleet? In his excellent book *Is Anyone Out There?*, Frank Drake makes his position clear: 'If I believed the ETs would come out of the sky, I wouldn't bother. I would just sit outside in a lawn chair and wait for them to show up.' I am inclined to agree.

Far more likely than any invasion is contact by radio signal. In that case, my hunch is that, once the existence of extraterrestrial intelligence is confirmed, it would then quickly settle into the background of everyday life. After all, aliens have long been a fixture of our entertainment culture – we are primed to accept their existence. In addition, the news cycle relating to any 'conversation' would simply be too long for it to capture the popular imagination even sporadically. TRAPPIST–1, for example, is just over forty light-years away and a message sent there today will not receive a reply for eighty years – more like opening a letter from your grandmother's pen pal than having a lively chat.

For decades, the value of SETI has sparked lively debate among scientists, politicians, policymakers, journalists. Yet, despite harsh budget cuts and scepticism, SETI persists. Our longing to answer the question of whether we are alone in the Universe burns so brightly in some that they dedicate their lives to the search, fully aware that only future generations may uncover the truth.

The search for extraterrestrial intelligence has not hinged on *whether* there is life is out there for a long time now. The Drake equation formulated what is needed for widespread life, and over the past decades astronomers have shown one by one that these factors are satisfied. There are too many galaxies, with too many stars, each with too many exoplanets, for us to think that none of them nurture life. When we look around the interstellar medium, the ingredients are all there, and will also be within these alien star systems, ready for the courier comets to mix them, scoop them and deposit them on any nearby planet. Yes, there will be many planets where life does not evolve, or planets where life suffers extinction events due to variable stars throwing out too much radiation. But there will also be plenty of planets around stable stars like our Sun, that sit just where liquid water can form. SETI hinges only on *where* we need to look, amongst the 400 billion or so stars in the Milky Way. A comprehensive search that large is going to take serious money.

But SETI has always found a way to survive, and in the last decade, a new generation of tech billionaires has sparked a second space age. Between 2019 and 2025, over 7,000 Starlink satellites were launched into low-Earth orbit, well on the way towards the 42,000 Space X has planned. Elon Musk's Space X and Jeff Bezos' Blue Origin are leaving national space agencies in the dust when it comes to developing new modes of reusable and cheap space travel. We are moving towards establishing a permanently crewed base on or around the Moon in a matter of years, and one on Mars in a matter of decades. Like our imagined scene within the TRAPPIST-1 system, humanity will soon be a species living across the planets, and travel between them will become possible for the moderately wealthy. Perhaps because we will soon be living extraterrestrial lives ourselves, there will also be renewed interest in what is Out There, and private funding to go with it. With the money at these space-obsessed billionaires' disposal, that monumental search catalogue of stars starts to look a little more achievable.

Whatever form that life takes, we know how we can expect to be contacted, because the laws of physics do not vary across the galaxy. Whether that planet has a yellow-green sky, whether it has five moons, or its people five legs, radio waves will be the form of light that is used for long-distance communication.

Perhaps, forty light-years away and forty years ago, a little green person on TRAPPIST-1e switched on a shiny radio telescope and began to send messages towards the star systems all around, crossing all twenty fingers that someone out there would hear them. Perhaps one day we'll turn our dishes towards TRAPPIST-1, and tune in just as that alien message reaches us. It could be in a hundred years. It could be tomorrow. Until that fateful day, we look at the stars, we wait and we listen.

OUR UNIVERSE

13

FIRST LIGHT

> There is stardust in your veins. We are literally, ultimately children of the stars.
>
> Jocelyn Bell Burnell, Euroscience Open Forum, Dublin 2012

Having examined the black hole at our galaxy's centre, tested general relativity with pulsar networks and searched nearby planets for alien signals, we now step back to take in the grander story of space and time. In this final part, we'll explore the cosmology of the Universe: from the forgotten first stars to the invisible scaffolding of dark matter, and finally to the dynamic, haunting presence of dark energy. First, though, we need to begin with how it all began: the Big Bang.

In 1964, astronomers Arno Penzias and Robert Wilson had a problem. Not with their science, at least not yet, but with two stubborn pigeons nesting in their radio telescope.

Their instrument, a vast, squashed-ice-cream-cone of a dish, was designed to shield against unwanted ground signals. While external interference was reflected away, a 6-m^2 opening at one end allowed signals from the sky to pour in, funnelling them towards the waiting antenna at the narrow end.[1]

Inside this metal cave, the air was still and warm and highly appealing, it turned out, to pigeons. The birds had settled comfortably near the antenna, roosting atop the warm electronics, blissfully unaware that their presence was about to collide with a Nobel-worthy discovery.

Penzias and Wilson wanted to use the antenna to study hydrogen in the Milky Way at its characteristic wavelength of 21 cm (1,420 MHz). As careful scientists, they first ran a null test at 4,080 MHz (7.3 cm), a frequency where they expected nothing, to ensure their equipment wasn't generating rogue signals. But instead of silence, they picked up an unexplained background hiss, eerily reminiscent of Karl Jansky's cosmic radio waves first detected at the very same Bell Labs facility three decades earlier.

Suspecting human interference, they aimed the telescope at New York City. No change. And, unlike the radio noise discovered by Jansky, which followed the constellation of Sagittarius, this signal came from everywhere, equally intense in all directions. Perhaps, they thought, the fault lay in the electronics. They painstakingly adjusted components, checked and rechecked connections. Still, the noise persisted.

With all logical sources exhausted, Penzias and Wilson turned to the pigeons. Or, more precisely, to what the pigeons had left behind. The birds had generously coated the inside of the telescope with what the astronomers delicately described in a scientific paper as a 'white dielectric material'. Could pigeon poop be guilty of skewing their readings?

Calling in a pigeon fancier, they had the birds relocated and after having the antenna scrubbed clean, they resumed their work.[2]

But these pigeons weren't any old pigeons. They were homing pigeons. Released by their new caretaker, they promptly flapped their way back to the telescope, settled in and resumed their business.

Faced with seeing their scientific progress stalled once again, Penzias and Wilson reluctantly resorted to a shotgun. 'So it's not something I'm happy about,' Penzias later admitted, 'but that seemed like the right thing. It seemed like the only way out of our dilemma.'[3]

Yet even after the pigeons were gone and the telescope was scrubbed clean once more, the signal remained. Wherever they pointed the antenna, the strange background noise lingered, with a temperature of precisely 2.7 K (about −270 °C).* This was no avian anomaly. By this point, radio astronomers were well accustomed to detecting signals from stars, galaxies, black holes and planets. But this was different. This was everywhere, as if space itself were bathed in it.

Then came a chance phone call.

On the other end was a team of Princeton astronomers, who had been painstakingly working through a bold prediction: if the Universe had truly begun with a vast explosion, then the relic radiation of that event should still be detectable, an all-pervading afterglow cooled to just a few degrees above absolute zero. Their calculations suggested it would appear as a signal across multiple frequencies, including 4,080 MHz. They had even begun constructing a radio telescope to hunt for it.[4]

But in a stroke of serendipity, Penzias and Wilson had beaten them to it.[5] All they had intended was a routine null test before beginning their study of the Milky Way. Yet after a maddening search for the source of this mysterious signal, and the small bang

* Kelvin (K) is the physicist's unit of temperature, scaled such that 0 K (about −273 °C) is 'absolute zero', the coldest temperature possible according to the laws of physics.

that marked the end of two stubborn pigeons, they realised that they had uncovered the faint, ancient echo of the Big Bang itself.

The Big Bang. It couldn't sound more dramatic if it tried, could it? Those three syllables encapsulate an idea so mind-boggling it defies intuition: that everything, the entire contents of the Universe, was once condensed into a point of effectively infinite density.[6] We don't know how long for, and it doesn't make sense to ask, as there was no time just as there was no space until, for reasons we still cannot begin to guess, spacetime itself began to expand.

The energies involved were immense, the expansion fast and the conditions hot and violent. This was no gentle unfolding of the Universe, like the turning page of a pop-up book, with fully formed stars and atoms popping into existence. Only the most simple, subatomic particles could endure the chaos. The Universe had to grow, cool down and earn its complexity.

The idea that the Universe was once radically different from how we see it now, that it evolved from a state of pure simplicity into a cosmos teeming with galaxies, planets and life, is a difficult one to accept. Look up at the night sky, and, aside from the motion induced by Earth's rotation, everything seems frozen in place. The stars of Orion remain in formation, as do all the constellations. The great Andromeda Galaxy can always be found between Cassiopeia and Pegasus and will look the same as it did a century ago. From night to night, year to year, we see no obvious change, no grand transformations in the sky. A person without access to a large telescope could be forgiven for assuming the Universe has always been as it is now. But in the 1920s, Edwin Hubble shattered that illusion and left his fellow astronomers reeling.[7]

Hubble studied the light from galaxies beyond our own and discovered something extraordinary: almost without exception, their light was redshifted, stretched towards longer wavelengths. Had the galaxies been free to move around space, he would have expected to see an equal number of galaxies red- and blue-shifted. His observations instead implied that almost all galaxies were moving away from us.* In addition, the expansion of space itself causes the furthest galaxies to appear to move away from us more rapidly: a relationship known as Hubble's law.

Line up any three small objects in front of you, each a hand's-width apart. From left to right I'm looking at a paperclip, a pen and a small Spock action figurine, but anything will do as we are just using them to represent galaxies. Let's say that, in the time since you laid your objects out, spacetime has expanded such that distances in space have doubled. From the perspective of my paperclip, my pen has moved along the line so it is now two hand's-widths away, which would land it on top of Spock. But space has expanded everywhere, and so I need to move Spock at the same time. He will now sit two hand's-widths away from the pen and four from the paperclip. From the paperclip's viewpoint, the pen moves one handspan, and Spock moves two handspans during that same moment of expansion, and so Spock appears to the paperclip to be moving faster.

If distances double, then the furthest object appears to move away twice as fast as the nearer one. If distances quadruple, the furthest object (Spock) would recede four times as fast.

Things that are held together by gravity or by the forces inside atoms don't get pulled apart by cosmic expansion – Spock stays whole. The stretch shows up only across the vast emptiness between galaxies, and even there it's incredibly gentle from

* A handful of nearby galaxies, including our neighbour Andromeda, are exceptions: their gravity pulls them towards us faster than cosmic expansion pushes them away. Over small distances, gravity wins; over large distances, expansion dominates. In roughly 4 billion years, Andromeda will merge with the Milky Way, forming one giant elliptical galaxy.

moment to moment; it only becomes noticeable when you add up billions of light-years and billions of years. But by carefully measuring how fast galaxies close to us are travelling compared to galaxies further away, astronomers can work out how quickly space is expanding over time.

Hubble's observations in 1929 dealt a serious blow to the dominant theory of the time, the idea of a static Universe that was at rest, with no beginning and no end. In the early twentieth century, Alexander Friedmann had worked with Einstein's equations of general relativity and found that an expanding (or contracting) Universe was mathematically possible. Einstein himself was appalled by the idea and added a 'cosmological constant' to his equations to keep the Universe static (an act he would later call his 'biggest blunder').[8]

Many astronomers clung to this idea too, unwilling to accept a Universe that was changing around them. Eventually though, as galaxy after galaxy was observed travelling outwards, the idea became the accepted cosmological dogma.

During the 1930s and 1940s, a handful of astronomers began to propose the idea of a primeval atom, an early version of what would become the Big Bang theory.[9] The backlash was fierce. In response, another group put forward an alternative: the Steady State theory. This new proposition allowed for an expanding Universe but rejected the idea of all matter once being compressed into a single point. Instead, they argued that as the Universe expanded, new hydrogen atoms would continuously appear, enabling galaxies to form indefinitely and maintaining a perfect cosmic balance. In this view, the Universe had no true origin, only perpetual renewal. One might imagine it as a fish tank that keeps growing while new fish are continually added, so the number of fish in any given space remains the same. The galaxies we see today may not be the same ones that existed billions of years ago, but on the largest scales, the nature of the Universe remained unchanged. It was expanding, they admitted, but they insisted that it had always done so – and always would.

The arguments between the two camps, Steady State and Big Bang, were intense, legendary even, playing out both in academic literature, lectures and the media.[10] The term 'Big Bang' itself was introduced by one of the Steady State theory's most loyal scientists: Fred Hoyle. In a 1949 BBC radio broadcast he used the term in a dismissive manner meant to ridicule the idea of a 'primeval atom', but the phrase stuck. For decades, the two theories tussled, until radio astronomy dealt the Steady State theory a triple blow.

Up until the mid-twentieth century, Edwin Hubble and his contemporaries primarily observed nearby galaxies. This was because the ground-based optical telescopes of the era faced atmospheric interference, which absorbed and scattered faint light from more distant galaxies. Radio waves had a far easier journey to Earth, gliding through the atmosphere with minimal interference. And the new generation of radio astronomers were about to reveal something even more profound: the Universe was filled with 'radio-loud' galaxies. These galaxies, which we now know are powered by supermassive black holes at their cores, blasted out immense amounts of radio waves, so much so that their radio emission entirely outshone their optical light. Not only could radio telescopes see further than optical telescopes from Earth, but the galaxies they were trying to see were naturally brighter in the radio too. Radio telescopes could tune in to these objects at staggering distances, completely beyond the reach of optical telescopes. And so, after the war, enthusiastic physicists began to catalogue the 'radio sky' – writing down the coordinates, brightness, shape and variability of every source of radio waves they observed.

Because optical astronomers had mapped the nearest galaxies, which are all a similar age, they had found no significant differences between them, reinforcing the Steady State theory. But radio astronomers had a different view. With their telescopes sensitive to far greater distances, they could catalogue galaxies at progressively earlier epochs in the Universe's history. And almost immediately, they noticed something odd. The further back in time they looked, the more galaxies they counted, even after adjusting for the larger amount of space they were surveying. They had uncovered what we now know as 'hierarchical galaxy evolution', where numerous small galaxies merge into fewer larger galaxies over billions of years.

But what really baffled the radio astronomers was that some of the brightest radio sources had no obvious galaxy associated with them and appeared point-like, almost star-like. And their spectra made no sense. All chemical elements absorb light at preferred wavelengths, such that their presence leaves predictable gaps. The observed light from these sources did not match any of the known spectral 'bar codes' of Earthly elements, suggesting the discovery of an unknown element.

The breakthrough came when astronomers realised that these spectral lines weren't exotic at all; they were the signatures of hydrogen and other familiar elements, just redshifted to wavelengths far beyond where anyone expected to see them.[11] Applying Hubble's Law, they found that these objects lay billions of light-years away. Our understanding of the sheer scale of outer space had exploded, almost overnight. Radio astronomy had redefined cosmology once again: the fish-tank had become an ocean.

These quasi-stellar radio sources, or *quasars*, were unlike anything astronomers had ever seen.[12] Given their extreme distances, their brightness was staggering. They were more luminous than entire galaxies, yet their light flickered on timescales of months, meaning they had to be incredibly compact. When trying to understand the rhythmic radio beats of a pulsar, astronomers used a simple rule: the timescale over which an object's light varies tells us something about its size. Big objects don't flicker quickly. Light from the edge has a longer journey to us than light from the middle, so the signal gets smeared out. Quasars varied on timescales of months, which was lengthy compared to pulsars but still a revelation. Despite their staggering brightness, the flickering meant that the radio light must be coming from a region no larger than our Solar System.

It could only be one thing: a black hole. Our own black hole is 'supermassive', but remember that massive doesn't mean big. Sagittarius A* would fit well within the orbit of Mercury, yet it has the mass of 4 million suns. In radio-loud galaxies, supermassive black holes devour surrounding matter; enormous energy is released from the hot material around them and in twin jets. What we call this phenomenon depends on perspective.[13] Viewed side-on, the jets inflate huge radio lobes, and we see a radio galaxy. If one of those jets happens to point almost straight at us, it is an extreme case called a blazar. More generally, when the wider radio lobe is pointing in our direction, we call it a quasar. Over time, the gas reserves around these black holes deplete, cutting off the bright beams of radio light, such that quasars become rarer closer to modern times.

The implications were profound. Quasars existed at tremendous distances, meaning they were a feature of the Universe billions of years ago. If the Universe had always looked the same, as Steady State theorists claimed, why had astronomers never found anything like them nearby? They certainly weren't hard to spot, after all. Their existence reinforced what the radio galaxy counts had already suggested: the Universe was changing, evolving, growing.

Yet still, the proponents of Steady State theory held firm. After all, what was this new-fangled radio telescope technology anyway? Nothing but wire and wood, and a breed of astronomer that didn't even exist ten years ago. It was easy to dismiss, especially when the inertia involved in shifting a long-held belief is so strong.

The deciding blow that knocked out the Steady State theory and the idea of a static, unchanging Universe once and for all was the detection of the afterglow of the Big Bang, the Cosmic Microwave Background (CMB), in 1964 by Penzias and Wilson.

In the moments following the Big Bang, the Universe was a seething broth of photons, quarks and electrons. A few millionths of a second after the Big Bang, these quarks came together to form protons and neutrons, and by about fourteen seconds after the Big Bang, the Universe had cooled enough for these protons and neutrons to combine into nuclei – the cores of atoms. All the nuclei and electrons and photons then existed in a state of thermal equilibrium, which meant that although particles were constantly colliding, forming and breaking apart, the overall mix stayed largely the same.

Quickly, though, the energies involved began to drop further. This gave the newly formed nuclei a chance to survive a little longer. Hydrogen and helium nuclei began to build up, but anything heavier was too fragile to hold together. Even a weak photon could smash them apart.

By the time three minutes and forty-six seconds had passed, this brief, fiery burst of creation had finished.[14] About a quarter of the Universe's ordinary mass had formed helium−4 nuclei, made of two protons and two neutrons. The rest was mostly hydrogen nuclei (single protons), with tiny traces of deuterium and a dusting of lithium. In the time it takes to poach an egg, the Universe had quietly stitched together the ingredients needed for the first stars, which would emerge hundreds of millions of years later.

But the work wasn't over. For hundreds of thousands of years, the Universe remained a hot, chaotic soup of nuclei, electrons and photons. Light couldn't travel freely. Instead, photons constantly ricocheted off electrons, their paths zigzagging wildly. If you've ever noticed the shimmer of heat above a road in summer, you've seen how varying air temperature can bend light. In the early Universe, the effect was far more intense. Light could never make it very far before it was scattered entirely, which is why we can't see back to these earliest moments with ordinary telescopes: it's like trying to see to the bottom of a bubbling jacuzzi.

For the next 380,000 years, the Universe remained an opaque glowing fog of charged particles until finally temperatures dropped far enough that electrons combined with the hydrogen and helium nuclei to form whole atoms: an event known as recombination. With far fewer electrons in the way, the photons could at last travel freely. The Universe, for the first time, became transparent: the jacuzzi was switched off.

Everywhere, there was light, an afterglow of the Big Bang, which would cool over time in a predictable fashion . . . if one assumed the Universe was expanding.

Though now measuring just 2.7 K, the CMB remains largely unchanged since the moment the Universe turned transparent – a faint afterglow, all around us, waiting for an antenna to pick it up. In the days of terrestrial analogue TV, around 1–2 per cent of the static that appeared when tuning in to an unused channel was caused by CMB radiation. Your own TV aerial allowed you to see the fallout from the birth of the cosmos, in between cartoons.

As this same relic radiation drifted serenely into the pigeon-free horn antenna in 1964, Penzias and Wilson unwittingly confirmed the existence of the CMB as predicted by Big Bang theory. When combined with the growing evidence from radio galaxy counts and quasars, the case against the Steady State theory became undeniable. Faced with such precise and compelling data, its proponents had no choice but to concede. The cosmological crown belonged to the Big Bang.

★

The Universe had a beginning, and a hot, violent one at that, which means everything within it had a beginning too. No stars or galaxies emerged fully formed from the Big Bang, and no elements heavier than hydrogen, helium and a trace of lithium could have formed without a little help from stellar fusion engines. If the Big Bang was the moment of conception and the CMB an early ultrasound image, then what follows is the story of the Universe's first steps.

If we could step into a time machine and travel back to just a hundred million years or so after the Big Bang, we'd find ourselves in an all-encompassing, suffocating blackness. The hydrogen and helium gas are invisible to our eyes, and even the glow of the CMB has stretched out of the visible and into the infrared. There is not even a flicker of light for our eyes to cling to, but beneath this darkness, an invisible scaffolding is already shaping the Universe, coaxing the hydrogen into filaments and nodes. Eventually, a lucky cloud of gas becomes dense enough to collapse, igniting fusion at its core, releasing light across the wavelengths and marking the birth of the very first star.

This star was unlike anything we see today. Stars today are mostly made up of hydrogen and helium, but they have a significant sprinkling of heavier elements too. The first stars were primordial stars, formed in a pristine Universe, comprising only those two lightest elements. This different chemical composition meant that they were hundreds of times the mass of our Sun, burning blue-hot from its intense temperature. All over the Universe, one by one, more stars ignited, filling the sky like fireflies and, as they did so, changing the Universe.

The Era of the First Stars and the Dark Ages that preceded it are my main research areas and the topic of my first book.[15] On the first day of my PhD, I was assigned my project: simulating optical observations of galaxies to see how a future telescope might help us pin down the properties of dark energy. There was a long pause as I took it in, until my supervisor finally noted, 'You don't look very happy.'

If you promise to forget I said this when reading Chapter 15, I can share a secret: I don't really like dark energy. I believe the science that supports its existence, but the concept itself feels too intangible to spark that small thrill of wonder – the kind I get from seeing a planetary nebula glowing in exquisite colour or watching a black hole revealed by its glorious radio jets.

Perhaps sensing my hesitation, I was given a second option – a small side project: observing the first stars ever to exist, using radio telescopes to look back in time. That was it. There was that thrill. A spark of curiosity that lit up a hundred questions, despite the fact I had never once thought about the first stars or even set eyes on a radio telescope. What were the first stars like? Was the Universe truly dark before them? Do any still survive? What about the first black holes? And how can we possibly study them?

I have spent over a decade searching for light from this time using radio telescopes, so that we can better understand the first billion years or so of the Universe's history, a time full of firsts.[16] It is a time of great uncertainty, because, with the exception of the CMB, observations within that time are incredibly difficult. We are gathering light that has travelled for 13 billion years or more, and so will have been redshifted to the extreme, and scattered and mixed up along the way.

It is important to fill in any information we can about this lost era, for several reasons. Firstly, we cannot hope to understand how we went from the Big Bang to the Universe we see around us today without access to those first billion years. This was when the first stars, first black holes, first galaxies, first supernovae all emerged. To have no knowledge of this formative time is equivalent to missing all a child's firsts: the first steps, the first word, the first wobbly tooth.

Secondly, for stars, big doesn't just mean beautiful, it means brief, as well. Their huge masses created so much gravitational pressure that they used up their fusion fuel of hydrogen in just a million years. Then, as suddenly as the first stars appeared, they exploded in brilliant supernovae, scattering their newly forged heavier elements across the cosmos, seeding it with the raw materials needed to create stars, planets and life.

Their vast masses, and short lives, mean the first stars are likely an extinct species, a type that cannot form in the present day. The Universe began to take on a yellower hue, as the destruction of the first stars paved the way for the next generation of stars, enriched with heavier elements. As if improving on a prototype, these stars were smaller, cooler, longer-lived, less blue and more yellow. Many still exist today, but most stars, including our Sun, are of the third generation, the youngest, containing the highest proportion of metals. The Universe is simply too polluted with heavy elements for any clouds of pure hydrogen and helium to exist any more.[17]

There is a small chance that, alongside the massive beasts that defined the first generation, a few smaller sibling stars formed too. They might have survived, gathering heavy elements from their surroundings and remaining camouflaged until the present day. One day we might have the technology to work out whether there are tiny first stars hidden in plain sight, but there is no doubt that to understand the Era of the First Stars, we need to look back in time.

As we've seen, one way to do this is to gather light that has travelled from far, far away by tuning in to its stretched wavelengths. The JWST has produced incredible images of galaxies billions of light-years away.[18] But individual stars are too small to detect using this method, and the very first galaxies are beyond even its extensive reach.

Indeed, when it comes to observing the first stars individually, even radio astronomy cannot come to the rescue. The light emitted from these stars, dazzling up close, is too faint to be detected at any wavelength, having travelled for over 13 billion years. We have never seen the blue light of a first star and, most likely, never will. We can, however, using radio telescopes, detect the hydrogen which surrounds them, and learn about their characteristics much as we learn about the dinosaurs from their fossilised footprints.

Hydrogen comprises a single proton and a single electron, which interact to produce 21-cm radiation, as we have seen many times in this book. After the Big Bang, the baryonic (i.e. non-dark-matter) part of the Universe was mostly hydrogen and practically singing with this radio light, even throughout the Dark Ages when there was no visible light to be seen. There is a lot of hydrogen pervading the early Universe and so this light is much brighter than any individual first star, or even first black hole or first galaxy. Able to stand the test of time and the relentless expansion of space, it forms a bedrock upon which the early Universe can tell its story.

As the first stars and black holes formed, their radiation swept through the cosmos, *ionising* the hydrogen such that the electron and proton were separated, and could no longer interact to produce radio waves. As the first stars continued to produce bright, UV light, more hydrogen was ionised and that 21-cm chorus was enveloped in great expanding bubbles of silence. By scanning the sky for 21-cm radiation, the hope is that we can map these voids, creating images that look like a slice of Swiss cheese. The shape, size and number of the 'holes' all help us better understand the nature of those first celestial objects, acting as unique fingerprints for the first stars, black holes and galaxies. Stars produce clear, crisp, circular holes, while black holes create larger, fuzzier gaps in the radio maps.

To look back in time, we don't need a time machine because

the Universe has recorded its own home videos. But these recordings aren't in a format just any telescope can read. Trying to detect them with the JWST would be like attempting to play a vinyl record on a CD player. Their story is written in the longest wavelengths of light, stretched by over 13 billion years of cosmic expansion, and can only be decoded by radio telescopes and vast, sprawling networks of antennas at that.

The telescope I have been most involved with is the Square Kilometre Array.[19] The SKA ('ess-kay-ay') is set to be the largest and most powerful radio telescope ever built, with the potential to transform our understanding of the Universe. It will consist of two vast arrays of antennas, with 131,000 low-frequency antennas spread across the Australian outback (SKA-Low) and nearly 200 mid-frequency dishes in South Africa (SKA-Mid), reaching across distances of up to 150 km. Once complete, the SKA will have a total collecting area of one square kilometre, making it the most sensitive radio telescope in history.

The scale of SKA is staggering. The data transfer rate will be 100,000 times faster than current broadband speeds, and the optical fibre network connecting the antennas will be long enough to wrap around Earth twice. Supercomputers distributed across the globe will process the flood of data, archiving enough information each year to fill one million laptops. It will tune in to the faint radio whispers of the early Universe, mapping the cosmic web in unprecedented detail, detecting gravitational waves through pulsar timing and even searching for signs of extraterrestrial life. If previous generations of radio telescopes are anything to go by, the SKA will enjoy well over a fifty-year lifespan and have ample time to not only solve today's biggest mysteries but also answer questions we have yet to even ask. The SKA isn't just a telescope. It is a time machine, a cosmic palaeontologist and a pioneer of the Era of the First Stars.

It has taken years to produce the technology capable of eavesdropping on this ancient era, and there are years more of meticulous observations ahead of us. But, as the next generation of radio telescopes prepares to press record, we stand at the threshold of uncovering the Universe's earliest moments – an era lost to starlight but not to time.

When I chose to study the Universe's distant past, I became part of a global research community. With emails, zoom calls and software which allows live contributions to research papers from every continent at once, I collaborate with an impressively wide network of astronomers. We don't need an office. As cheesy as it sounds, Earth is our office. The SKA especially embodies this collaborative spirit. To build a telescope this ambitious requires expertise that cannot be found in one country alone, and so the team hails from over fifteen countries across five continents. It is a global observatory, pooling resources and knowledge, even attempting to distribute manufacturing and design contracts fairly amongst its members. The breadth of the SKA project is such that its members get together in a physical location only every ten years. I was at the conference dinner in Sicily in 2014, as we watched Mount Etna erupt into the night sky, when we hadn't even worked out what the telescope would look like. And by the time this book is published I will have attended another conference, this time deep in the Black Forest in Germany, and seen just how far our technology and understanding has advanced. As the years go by, my curiosity only increases, for all of it, all of the science this wonderful instrument is exploring.

And that's just one telescope. There are many more, dotted all over the planet, with different teams attached, all listening for the echoes of the first stars. From the Australian outback to the South African desert, from the rolling English countryside to the mudflats of the Netherlands, networks of radio antennas stand tall, listening for the faint whispers of the Universe's earliest moments. Every radio wave that has made it to Earth from the Dark Ages and the Era of the First Stars is a fragment of an ancient story that began with the Big Bang. But that story has a hidden author, dark matter, without which the plot doesn't make sense.

14

DARK MATTER

> In a spiral galaxy, the ratio of dark-to-light matter is about a factor of ten. That's probably a good number for the ratio of our ignorance to knowledge.
>
> Vera Rubin, Address to International Astronomical Union, 1985

In the last chapter, we looked back across cosmic time, from the Big Bang to the first stars. Now our view shifts: instead of peering into the past, we move outwards through space, leaving the Milky Way entirely and venturing into intergalactic space.

There is no clear border to cross as we zoom to the tips of the spiral arms, no 'Thank You for Visiting Our Galaxy, Please Drive Carefully' sign. Instead, the distribution of stars simply becomes sparser and sparser, like the towers of a great city blending into suburbs and then countryside. Eventually, we simply pass the last star and with it, maybe, the last planet.

If intelligent lifeforms are enjoying a life on that isolated outpost, their night sky would shift according to how their world turns and where they stand upon it. The same is true for us. I remember visiting Australia, where the sky was the most captivating tourist attraction of all. Orion, the familiar hunter, appeared jarringly upside down, and I saw the iconic Southern Cross constellation for the first time. The ribbon of the Milky Way was different too, as the tilt of Earth treats the southern hemisphere to a better view of its dense and bright heart.

But for those alien astronomers at the edge of the galaxy, the contrast would be even starker. Half of the sky would be ablaze with the full view of our galaxy. Those on the other side of that last planet, however, would find themselves staring out into a darkness punctuated only by the faint smudge of distant galaxies.

One might expect that, with no pesky stars getting in the way, a clear night sky might reveal a glittering sweep of galaxies instead. After all, in a dark location, you can see the smudge of our neighbouring galaxy Andromeda with the naked eye, and several more will pop out if you're armed with a basic pair of binoculars. There is no shortage of galaxies in the Universe, but the distance between them is so much larger than the size of the galaxy itself, that moving to the edge of the Milky Way does nothing to improve visibility. It's like climbing a ladder and expecting to get a better view of the Moon. It is tempting to yearn for a closer look at distant galaxies, to gather more light and unveil the intricacies of each individual one. But it is the invisible connections *between* the galaxies that can lead us to the most elusive cosmic truths.

This book began with our feet on solid ground, as we explored how radio astronomy has illuminated our Solar System. Our destinations, Venus, Mars and the Moon, for example, were tangible: we can look up and see them ourselves. In the second part of the book, whole constellations, such as Sagittarius, Taurus and Aquila, have served as a backdrop to our explorations. Now, in these final chapters, we delve into how radio astronomy can trace the hidden architecture between galaxies and shed light on the Universe's darkest secrets: dark matter and dark energy.

Both phenomena are theoretical constructs designed to explain peculiar behaviours observed in the Universe. Hints of dark matter go back to the 1930s, but the theory only gained traction in the 1970s, when studies of other galaxies confirmed that stars were orbiting faster than expected, implying missing matter. Dark energy, meanwhile, was proposed in the 1990s, when observations of distant supernovae revealed that the expansion of the Universe is not slowing down as expected, but accelerating. Both are quite recent theories, in the context of astronomy as an ancient science. Generations of astronomers have got through their careers without ever needing to resort to exotic theories of missing matter and energy. The effects of dark matter and dark energy only become apparent on the grandest scales, far beyond the reach of human senses, the scales of planets or even star systems. The kind of optical telescopes required to investigate these theories are behemoths, requiring huge collaborations of scientists, generous wallets and monumental patience for politics. They are undoubtedly powerful facilities, but there is a complementary, often more efficient, way to probe these scales.

Thanks to radio astronomy, we can explore these vast scales comparatively quickly and cheaply, albeit in a cruder fashion. I call it '8-bit astronomy', a nod to the pixellated computer games of the 1990s that consumed my childhood. Back then, game designers didn't have the technology required for realistic graphics, but that wasn't the point. The magic of those games lay in their stories and gameplay, their ability to draw you into a world, no matter how blocky the visuals. Today, we have the tools to create hyper-realistic game environments, driven by graphics engines that simulate real-world and real-time physics. And that's marvellous if you're after a pretty picture or want to play a game that feels like real life. But if it's the immersivity of a good story you're interested in, sometimes visual simplicity is key.

This philosophy is at the heart of radio astronomy. In many areas, radio telescopes have progressed such that they can, and do, produce gorgeous images of, say, galaxies, competitive with any the Hubble Space Telescope might produce. But often, instead of striving for high-resolution images comparable to what optical telescopes can give, radio astronomers strip away the glitter and graphics to get to the raw data that describes the whole story: reflections from buried ice, the scruff of a pulsar turning, a flare from the Sun. When it comes to cosmology, this '8-bit' approach isn't just a necessity due to the distances involved. It is the arena where radio astronomers are most at home, ready to dive straight into the heart of the game and unlock new levels of understanding about the mechanics of the Universe itself.

Every object, person, animal and plant you can see, have ever seen or ever will see, is made of 'normal' matter. Physicists call this baryonic matter. What makes this kind of matter visible, or at least detectable, is its ability to interact with light. It reflects, absorbs or scatters different wavelengths of it. Sometimes it is visible light, other times it is beyond our natural range. Particles of baryonic matter cannot help but play with light, and it always reveals itself one way or another in our instruments.

Imagine a pie chart representing the contents of our Universe. All baryonic matter, everything you or our best instruments have ever seen, accounts for a mere 5 per cent of that pie. It is worth repeating: we do not know what makes up 95 per cent of the Universe. All is not lost, however; we can at least divide our ignorance into two great portions of the pie, which we will consume one by one in this and the following chapter.

One slice of the pie is 'dark matter'. We refer to it as 'dark' matter because it does not interact with light of any kind, including radio light. Yet this unseen material is far from passive: it forms the invisible scaffolding I mentioned earlier, coaxing clouds of hydrogen into the first stars and galaxies. It's the ultimate stealth material, invisible to radar, X-rays, infrared goggles or the brightest spotlight. While dark matter might sound rare and exotic, it's quite the opposite. About 27 per cent of the pie represents 'dark matter', which means that, for every kilogram of 'normal/baryonic' matter, there are about six or seven kilograms of dark matter in the Universe. Not that we have ever seen a particle of dark matter, let alone captured or created one. And scientists have really tried.

So how do we know it exists at all? The key lies in gravity. Dark matter has mass, meaning it distorts spacetime and exerts a gravitational pull on the visible (baryonic) matter around it. Imagine an invisible man walking across a bouncy castle. You wouldn't see him, but you'd see the material sagging beneath his weight, and perhaps even find yourself tipping towards the indentation.

In the analogy of the bouncy castle, the sagging effect caused by the invisible man would be immediately noticeable to you because you and your invisible counterpart have a similar mass: you're both human-sized. You can feel the effect of his presence. Were the invisible man an invisible mouse, however, you'd be none the wiser even if it scampered next to you, because the deformations it caused would be too small to affect you.

Dark matter might as well be an invisible butterfly. While dark matter makes up over six times more of the Universe's total mass than baryonic matter, it cannot clump together into dense structures like stars, planets, humans. The way it behaves is fundamentally different from baryonic matter. When a large mass of baryonic matter, say an interstellar gas cloud, begins to collapse into a new star, it has to lose heat energy along the way – otherwise the cloud will just puff up again. It does this by emitting light, the perfect courier for energy. In contrast, dark matter lacks the ability to lose energy by emitting photons. We know this because, otherwise, we would be able to detect that light. Therefore when clouds of dark matter do collapse, and they do, the increasing pressure and heat prevents them from collapsing down far enough to create, say, dark matter stars. Thus, dark matter, despite being prevalent overall, is much more spread out in the Universe than normal matter.

To put this in perspective, the average human is a tightly packed collection of baryons, occupying about 70 l and weighing roughly 70 kg. By contrast, if that same volume were filled with dark matter at its average cosmic density, it would weigh less than one proton.

The challenge in detecting dark matter is that you only really notice it on galactic scales, where enormous clouds of the stuff outweigh the stars and gas we can see. In planetary systems like our own, most of the mass is contained in a central star. The Sun alone makes up 99.86 per cent of the Solar System's mass, and its weakening gravitational pull explains the motion of the planets as we look outwards. Mercury zips around the Sun in eighty-eight days at approximately 47 km/s, while the distant Neptune takes 165 years to complete an orbit at just 5 km/s. By the same logic, one would expect galaxies and even clusters of galaxies to behave the same way: the further out you go, the slower things should move. But that's not what we see.

Swiss-American astronomer Fritz Zwicky measured the motion of galaxies in the Coma cluster in the 1930s and found that they were moving far too quickly. The combined gravitational force of all the galaxies in the cluster, based on the visible mass, was not enough to prevent these galaxies flying out of the system, yet it stayed together. To account for the discrepancy, Zwicky proposed the existence of vast amounts of unseen mass, which he called *dunkle Materie*, or dark matter. Though he may have coined the name, his evidence was sketchy, and most of his peers set it aside. Only with Vera Rubin's comprehensive study of galaxy rotation curves, forty years later, did dark matter step out of speculation and into the mainstream.

By measuring the speeds of stars all the way out to the edges of spiral galaxies, Rubin found that they did not slow down as expected. Even when astronomers accounted for the combined mass of hundreds of billions of stars, the gas and the central black hole, they could not explain why the outer stars orbited just as fast as those near the centre. It was as if Mercury, Earth and Neptune all raced around the Sun at the same speed. The reason is that the speed of an orbiting body depends on *all* of the mass within its orbit, not just the big thing in the centre. For planetary systems, this detail is pretty much irrelevant because most of the mass is in the form of a central star anyway. But with galaxies, as you move further out, you begin to enclose huge volumes of dark matter whose gravity balances out the weakening influence of all the baryonic matter. The result was Rubin's flat rotation curves: galaxies behaving as though they are embedded in enormous halos of invisible mass. Her meticulous observations turned Zwicky's bold suggestion into a cornerstone of modern cosmology.

If Zwicky named the problem and Rubin eventually nailed the evidence, it was a Dutch astronomer named Jan Hendrik Oort who shaped the way astronomers would pursue it, by championing radio methods that could reach far beyond the limits of optical telescopes.

The year before Zwicky coined the idea of *dunkle Materie*, in 1932, Oort had noticed a discrepancy in the motions of stars close to the Sun. His calculations for the local mass density implied by those motions didn't add up, even when including all the mass of all the visible stars and interstellar gas. Looking back, this can be seen as one of the first hints that something was missing, though it was not yet recognised as the galaxy-scale dark matter we talk about today. It was a mystery that continued to niggle at him through the following years.

By the early 1940s, the Second World War was raging, Holland was under German control, and Oort had resigned from his academic position at Leiden University, in protest at the dismissal of all Jewish professors. He waited out the war in a tiny Dutch village in relative seclusion, but, unable to resist the pull of academia, he persisted in meeting colleagues for discussion, making the 100-km journey to Leiden by bike. Oort wanted to understand why the stars were moving faster than they should be, but optical measurements were stymied in many directions by obscuring dust. He needed to pierce through this dust and see further than optical telescopes would allow. But how? The answer came in reports of Karl Jansky's discovery, in 1933, that the hiss in his antenna was coming from the galactic centre. Oort realised that by switching from optical to radio, he could dispense with trying to see stars through the obscuring dust and observe the gas between them instead.

Even before the war ended, Oort tasked student Hendrik Christoffel van de Hulst with searching for a way to detect and map the invisible interstellar gas. Van de Hulst identified 21-cm radiation, emitted by hydrogen, as a promising candidate.

Oort wanted to pursue van de Hulst's theories immediately, but his reduced circumstances in the Netherlands hindered him, even after the war ended in 1945. Across the globe, physicists who had been tasked with developing wartime radar equipment began turning their skills and devices up to the skies, establishing radio astronomy as an academic research field. In contrast to the circumstances of astronomers in the US, UK and Australia, for example, much of Holland had been destroyed, and there were no radar groups with in-house expertise to tap. Oort was an old-school astronomer who knew the optical skies like the back of his hand, but the electrical engineering knowledge required to construct an interstellar listening device was a unique skill, and one that he lacked. Nonetheless, substituting enthusiasm for expertise, van de Hulst and Oort salvaged two leftover German antennas and set to work. The pace was slow, and the mistakes were many, but they made steady progress, working hard to identify and subdue all the terrestrial sources of noise that initially overwhelmed their data, burying a galactic 21-cm signal they weren't even sure was there.

Unbeknownst to them, two Harvard physicists had heard about the promise of 21-cm radiation and, unlike the Dutch team, also had the experimental expertise needed to follow the idea up. Edward Purcell was interested because of his background in identifying atomic structures by detecting the specific wavelengths of light the element emitted. Harold 'Doc' Ewen was a graduate student with a wartime background in radar physics, in search of a thesis topic to research alongside his full-time job at Harvard's Nuclear Laboratory.

Purcell was able to secure a small amount of funding and spent $800 on plywood, copper plating and the electronics to construct a 'pyramidal horn' antenna about the length of a car. This kind of telescope, shaped like a squared-off ice-cream cone, is less susceptible to interference. If a wide-open traditional dish is an ear, this telescope is like a cupped hand around that ear, pointed directly at the sky.

They fixed the tapered end of the antenna to an open window in a fourth-floor lab on campus, erecting scaffolding to support the wider 1.4 x 1 m opening at the other end. This allowed them to sit in the warmth of the laboratory, safe from the elements. Well, almost safe. Unfortunately, they did not foresee the risk of rain funnelling down into the lab, or that the opening of the antenna would act as an irresistible target for snowballs thrown by students.

Progress was gradual, but only because Ewen had to fit all the work around his job, working through weekends and holidays. On Easter morning in March 1951, Ewen's first spare day from work since Purcell had funded $300 of improvements from his own wallet, they successfully detected the 21-cm radiation from the Milky Way.[1] To optical eyes, it appeared they were pointing at nothing but empty space, and in the bright daylight. But van de Hulst had been right: radio telescopes could 'see' the invisible interstellar gas.

In a complete coincidence, van de Hulst was visiting Harvard at the time of the discovery, in order to teach a course on radio astronomy. He heard of Ewen's success shortly after his arrival and relayed the news to Oort, who, it turned out, was not having the best time without him. He had just lost years of painstaking work when the building housing the electronics had burned down, without a clear cause. It is perhaps not surprising that an electrical fire broke out. This was, after all, a group of astronomers playing at being radio engineers and fumbling with electrical circuitry they did not fully understand. In comparison, Purcell and Ewen were a lab-based physicist and radio engineer respectively, playing at being astronomers.

A rapid exchange of knowledge began, with the Americans teaching the Dutch how to construct the experiment, and the Dutch explaining to the Americans the astronomical context of what they had achieved. As a result, Oort achieved a detection only seven weeks later than Ewen, and the news spread to other international groups. This generous sharing of knowledge meant that, within the space of three months, the global radio astronomy community was equipped to see the invisible gas pervading our galaxy. By becoming a radio astronomer, Oort had the means to further investigate the niggling problem presented by his optical measurements twenty years earlier.

★

Looking at an optical image of a galaxy, we naturally define its border by the sudden lack of starlight. But observing in the radio wavelengths, we see that border expand dramatically, revealing a nebulous distribution of hydrogen stretching twice or even three times as far as the stars.[2] The first measurements hinting at dark matter measured the velocities of galaxies, and then the stars within them. Within galaxies, this method performs beautifully in dense, star-packed regions but becomes much more unreliable the further out from the centre of the galaxy you look. Radio telescopes, however, can measure the movement of the gas near the centre, through the region where the stars are too far away for their light to reach us, and far past where there are no stars at all.

The gas in a galaxy rotates just as the stars do. From our perspective, some hydrogen atoms will be moving toward Earth, while others recede. This motion introduces a Doppler shift in the 21-cm radiation emitted by the gas, with one side appearing slightly 'bluer' (shorter wavelengths) and the other slightly 'redder' (longer wavelengths). These subtle shifts allow us to map the speed and direction of the rotating gas and infer the mass of the galaxy, in an independent measure to the speed of the stars.

In 1957, after Zwicky's measurements, but a decade before Rubin's, van de Hulst led a team that measured the rotation of the Andromeda Galaxy using radio waves emitted by the gas. The gas raced around at breakneck speed right up to the point where even the gas petered out.[3] This can only mean one thing: the presence of an unseen mass – dark matter – stretching far beyond the galaxy's optical limits. At the time, though, radio results alone weren't enough to settle the debate. In the 1950s, radio astronomy was still a young science, and the breakthrough had to wait for Rubin, who proved that the stars told the same story. In the decades since, using both optical and radio measurements, astronomers have confirmed that this phenomenon is not unique to the Milky Way, or Andromeda or the Coma cluster; it is universal.[4] Dark matter is everywhere.

It is surprising that, forty-five years later, we still have to refer to the mysterious substance as 'dark matter'. The evidence that something is there is unequivocal, but what is there is still very much up for debate. Some physicists have been merrily colliding atoms at speeds approaching that of light to try and create the elusive particle, while others are searching for evidence of their passage through massive underground vats of dry-cleaning liquid. There have been no direct detections yet, but indirect evidence from both optical and radio observations has allowed us to better understand its nature, and its vital importance for explaining why the Universe looks as it does.

Astronomers have run many simulations, with differing levels and properties of dark matter, comparing the virtual universes created with the galaxies we see today, in order to work out how this hidden mass behaves. Together with the data, the simulations suggest that after the Big Bang, dark matter began to condense into *filaments*, forming *nodes* where these filaments intersected. While dark matter accumulated at these gravitationally intense nodes, it could not collapse into anything denser than diffuse clouds.

This web of dark matter made up the invisible cosmic framework upon which baryonic matter then settled. The gravitational pull of dark matter channelled hydrogen gas along the filaments and into the nodes.[5] But because baryonic gas can dissipate the heat associated with collapse in the form of photons, the hydrogen atoms could fuse to form the first stars. Soon, the combined gravitational pull of hundreds of stars and the surrounding gas formed the first galaxies. The clouds of incompressible dark matter still surround the galaxies and galaxy clusters at these nodes, and we call them *halos*.

It is important to remember that any beautiful optical image of a galaxy has been heavily filtered, containing only the optical light from the stars. With the radio data included, the galaxy's boundaries extend by up to double, and sometimes even more. With the dark matter . . . well, we haven't worked out a way to capture an image of that yet, but we can use the movement of the stars and then, far past where the optical light fades, the movement of the gas to sketch out our own image of a massive halo extending even further out.

To observe the diffuse gas, we have to zoom out, lowering our resolution. We swap the crisp spirals and glittering stars for hazy clouds that look like experimental watercolour paintings at best and pixellated JPEGs at worst. But what these images lack in sharpness, they make up for in reach and meaning.

It is a scientific trade-off: a shift in focus from the graphics to the story; from the decorative to the foundational. Zooming out, letting go of the stars, we begin to see the true shape of a galaxy complete with the expansive smudge of gas. Elegant? Perhaps not. Enlightening? Certainly.

And so, though some questions remain, we have now explored two sections of the pie chart making up our Universe: 5 per cent normal matter, 27 per cent dark matter. At this point, we might feel a bit smug, having figured out how to use radio waves to probe the distribution of dark matter over large distances. We have uncovered the scaffolding of the Universe! But one cannot help but notice that there is a full two-thirds of the pie taken up by something called 'dark energy'.

Despite their similar names, dark matter and dark energy are not related in terms of physics. It is an unfortunate nomenclature that often leads to confusion. Dark matter is a hidden form of matter that exerts gravitational influence, shaping the structure of galaxies and the Universe itself. Dark energy, on the other hand, is an entirely different phenomenon: a mysterious force driving the accelerated expansion of the Universe. Physicists have labelled it 'dark' not because it is related to dark matter, but simply because it, too, eludes direct observation. We cannot see it, we cannot yet understand it. With both concepts, we are in the dark. But if you thought this chapter was full of mystery, you'd better brace yourself, because dark energy makes dark matter look as common as mud.

15

DARK ENERGY

The missing link in cosmology is the nature of dark matter and dark energy.

Stephen Hawking,
Caltech lecture, April 2013

The galaxies we see today are like strings of fairy lights, draped along an invisible scaffold of dark matter that stretches across space. Radio telescopes, as we have seen, are expert at standing back and looking past the decoration to see the architecture beneath. In the previous chapter we zoomed past the optical boundaries of a galaxy two or three times, in order to measure the speed of the huge gas cloud enveloping the stellar part of the galaxy and so sketch out the dark matter halo. In this chapter, we need to think even bigger.

To get a sense of the scale of what lies ahead, consider this.

In Chapter 1, we began our journey by launching a radio photon from Earth. That photon took just 1.5 seconds to reach the Moon in Chapter 2. To get to the Sun, Venus and Mars, it needed only a few minutes. By Chapter 7, we'd arrived at Saturn, and the time had stretched to over an hour. There is a big leap in travel time to get to the nearest star, four light-years, and reaching the edge of the furthest star in the Milky Way would take light (*light*, the fastest thing there is) tens of thousands of years. To fully trace the presence of dark matter in a galaxy, we had to look much further out, reaching light travel times closer to a hundred thousand light-years. Yet even these staggering timescales pale in comparison to the 2.5 *million* years it would take that radio wave to reach our next-door-neighbour galaxy, Andromeda.

To begin to understand dark energy, we have to investigate galaxies so distant that through an optical telescope they appear not as grand spirals or majestic giants, but as dim, star-like points. To get to these galaxies, you'll need to stretch your imagination across distances so vast that our radio photon will take billions of years to traverse them.

Until 1998, the pie chart of the Universe's contents would have consisted only of what we have discussed so far: a greedy majority for dark matter, a slice for normal matter and a tiny sliver for all the photons knocking about. Astronomers enjoyed well over half a century of thinking that their pie chart was full of just these portions. It was quite a shock when they realised they had to make room for another, and a huge one at that. This discovery redefined the Universe for every astronomer at the millennium.

The first clue that we had miscut the pie came from exploding stars.

It is a fortunate quirk of physics that stars end their lives in predictable ways, based on their mass. The more massive stars collapse into black holes, the slightly less massive into neutron stars. These exotic endpoints, though much discussed, will be the fate of only a small fraction of stars in the Milky Way (well under a few per cent by number). The vast majority of stars don't have masses sufficient to create such extreme remnants. Instead, their collapse is halted earlier by electron degeneracy pressure, a quantum mechanical effect akin to trying to force two magnets to touch at their repelling poles. The resulting white dwarf, though not as compressed as a neutron star, is still incredibly dense; for example, the Sun will one day collapse into a white dwarf no larger than Earth and slowly fade from all visibility, over trillions of years.

About half of Sun-like stars have at least one stellar companion, most in binary systems. They're born together, live together, but rarely die together. The first star to form a white dwarf has the advantage, because it is able to siphon material from its neighbour, steadily gaining mass . . . at the expense of its sibling. The electron degeneracy pressure holds firm for a while, but eventually the increased pressure from all that extra gas becomes too much, triggering a runaway chain of nuclear fusion reactions. The white dwarf begins to enjoy a second lease of life, an alternative to that long descent into obscurity . . . but only for a few seconds. Almost immediately, the white dwarf explodes in a brilliantly bright 'Type Ia supernova'.

Type Ia supernovae are particularly useful to astronomers because their peak luminosity is remarkably uniform. To understand how we use them, imagine a friend holding a torch. If they stand just a metre away, the light is blinding, but if they move to the far side of a field, it's much dimmer. The bulb's luminosity (its intrinsic brightness) hasn't changed; the light is simply spreading over a larger area as your friend moves further away. Knowing the torch's luminosity, you can calculate how far away your friend is based on how bright (or dim) it appears.

In the same way, because we know the precise mass at which

Type Ia supernovae explode, we know how much fuel is behind the explosion, and so can estimate how bright they are up close. By measuring the brightness and comparing it to this estimation, we can work out how far away the supernovae are.

Type Ia supernovae are so bright that astronomers can use this method to estimate the distance to other galaxies with remarkable accuracy. We know the method works because there are other ways to measure cosmic distance, including redshift. Imagine that your torch-wielding friend is running away from you, simulating the expansion of the Universe. As they run, the light from their torch shifts to redder wavelengths: redshift, caused by the stretching of light as space expands.

In the late 1920s, Hubble had defined a neat, linear relation between a galaxy's distance and the speed at which it appears to recede, as shown by the redshift of its light. For example, 21-cm radiation emitted by a galaxy receding at 60,000 km/s would be observed as having a wavelength of 25 cm – a redshift of about 0.18, meaning the light's wavelength has been stretched by nearly 20 per cent. The neatness and predictability of relations like this are the reason I love physics. They have a comforting reliability: these laws apply everywhere and to everything.

Except when they don't.

Quite suddenly, our familiar understanding of our planet, the Solar System, the galaxy or even the Universe, can be shaken to its foundations by one observation. One observation that wakes us up from complacency and forces us to challenge a tiny assumption made in a theory so embedded that we had forgotten we had made any assumptions at all.

In 1998, astronomers studying Type Ia supernovae in other galaxies made just such a startling discovery.[1] They had two ways of working out how far away the supernovae were: comparing their brightness to the known intrinsic brightness of a Type Ia supernova and measuring the redshift of the galaxy. The two methods should have resulted in the same answer, but they didn't. Instead, the team observed each supernova to be dimmer than the redshift of the galaxy suggested it should be: as if your friend shouted that he was 10 m away but his torch was barely visible.

The distance determined from redshift rests on the Universe expanding at a steady rate, but these results suggested the galaxies were further away than we thought, and that the Universe's expansion was accelerating.

This discovery was deeply unsettling. The Big Bang theory already told us that spacetime is expanding, but astronomers had expected the expansion to slow down due to the gravitational pull of all the matter within. To learn that the expansion was speeding up was an enormous surprise, pointing to the existence of a mysterious, repulsive force pushing against that gravitational pull: a force we now call dark energy.

Dark energy seems to reveal its presence through its ability to speed up the expansion of space, so if we want to know more, we need to look at how the expansion of the Universe has changed over time. This might seem like a tall order, given that no one was around 10 billion years ago to make the measurements or keep the records. However, astronomers have a way of peering into the past. This ability stems from the finite speed of light, a fundamental property of the Universe that has little direct effect on our day-to-day lives and so can feel misleadingly magical.

Imagine your friend flashing their torch in Morse code. The message feels instantaneous because the distance between you is tiny compared to the speed of light, which travels at 300,000 km/s. We only begin to notice slight delays when sending messages between planets. But jump to our nearest galactic neighbour, the Andromeda Galaxy, and light takes 2.5 million years to travel here.

By exploiting how this time delay builds across huge distances, astronomers are able to look back at what was going on in the Universe billions of years ago. If we look at Andromeda, we see it as it was 2.5 million years ago. The distances are far, far too large to see supernovae, or even any individual star. Even galaxies appear just as smudges, in the radio or optical wavelengths. But we don't need the galaxies to look pretty, we just need to count them.

At the start of this chapter, I described the structure of the Universe as a scaffolding of dark matter upon which hangs a string of fairy lights, the galaxies. That scaffolding is formed of filaments and nodes of dark matter, with vast, empty voids in between. The filaments aren't lined up; it isn't a nicely ordered lattice. It is an improvised three-dimensional web spun by an idiosyncratic spider, with filaments crossing and passing at all angles. However, it is not complete chaos.

Telescopes have mapped the galaxies around us and traced the filaments of the hidden dark matter web. Amid the apparent disorder, the hint of a pattern emerges: on average, galaxies are ever so slightly more likely to cluster together at intervals of roughly 500 million light-years.[2]

This regularity is the fossil imprint of a cosmic drumbeat. In the early Universe, space itself rippled with waves caused by the volatile nature of gravity and radiation. These waves pulsed through the hot, dense plasma like vibrations in a drumskin. When the Universe cooled, those ancient beats gently guided the growing dark matter web, and some of the filaments came to align with the peaks of those waves. As the galaxies formed within those filaments, they echoed that preferred spacing of 500 million light-years.

It is certainly not the case that all filaments follow that fossil pattern, far from it. But with enough galaxies, the preference shows up, in the same way as a biased coin only reveals itself if you flip it enough times. The favoured spacing, woven deep into that tangled web, acts as a 'cosmic ruler' that we can use to measure the scale of the Universe across cosmic history.

Because light takes time to travel, when we look at distant galaxies we are also looking back in time. When astronomers repeat the same galaxy survey but focus on more distant galaxies, and therefore earlier times, they can measure the cosmic ruler again. In the younger Universe, the ruler was shorter, and cosmic expansion has stretched it since, so measuring how quickly or slowly that ruler grows lets us trace the history of the expanding Universe.

This gives us a method much like that of the supernovae: we know an intrinsic quantity and can compare that to what we measure, so as to work out how much spacetime has expanded in the intervening time. Now we just need to count all the galaxies from here back to the Big Bang . . .

In big optical surveys, astronomers observe the positions of galaxies in immense numbers, mapping their distributions across vast regions of space. But observing galaxies and stars across such vast spans of time is slow, error-prone work, and requires enormous datasets to reduce uncertainties.

Radio astronomers take a radically different approach. Surrounding those galaxies, across much bigger volumes, is hydrogen, emitting the distinctive 21-cm wavelength of radio waves. In order to study dark matter, we dispensed with the individual stars and instead concentrated on the hydrogen surrounding and pervading galaxies, and we called it 8-bit astronomy. In order to study dark energy, radio astronomers don't just dispense with the stars, they dispense with whole galaxies.[3] We might as well call it 4-bit astronomy.

The resolution in the resulting visuals may be terrible, but the story they tell is just as rich and, crucially, far more efficient. A single pixel of sky in a radio map might contain the gas from multiple galaxies, perhaps even tens of galaxies like our own. While this means we lose details like their shapes or colours, those details aren't needed. Just as you don't need to count every tree to map a forest, we don't need to identify every galaxy to detect a bias in their spacing. When we make that concession, radio surveys like this can be carried out in hours and weeks, not the months and years more detailed observations might take.[4] As I write, the first telescopes to apply this method are analysing their first results and, like archaeologists, piecing together the timeline of spacetime expansion.[5] Has dark energy always worked to speed it up? Has it ever pulled back the expansion, slowing it down? And what exactly *is* it?

Dark energy is the name we give to the mysterious force driving the accelerated expansion of the Universe. That's all we've got. You can see why radio astronomers want to hurry along our understanding.

Whatever dark energy is, its consequences are profound. If dark energy gets stronger, it could speed up the expansion so much that galaxies, then stars, and then even atoms are torn apart in a Big Rip.

If it remains constant, the Universe will expand for ever, growing colder and emptier in a Big Chill. Galaxies will drift apart, fading from each other's sights. As the stars naturally run out of fuel, they will fade, and even the black holes will eventually evaporate, over trillions of years. Then, the cosmos will be cold, dark and silent, empty of vibrancy, colour, life, for the rest of infinity.

If dark energy weakens, then gravity may one day gain the upper hand, as all the matter in the Universe exerts an overall inward gravitational pull, drawing spacetime back, condensing it once again. Galaxies would fly towards each other, then collide in a Big Crunch. Stars would be crushed into smaller spaces, temperatures would soar, chaos would reign, as the whole Universe collapsed into a hot dense state that might settle for ever – or trigger a second Big Bang, starting the whole Bang/Crunch cycle all over again, a new beginning for the Universe, though no life or structure will have survived the cataclysmic collapse.[6]

Cheery, eh? Whichever of these speculative scenarios plays out, it will be trillions of years in the future, so we shouldn't let it worry us too much. For now, we are in an era where dark energy dominates, and that is reason enough to pursue its true nature.

We do not understand 95 per cent of our Universe. The contributions from dark energy and dark matter remain invisible, betraying their presence only through their effects on the 5 per cent we do understand. Dark matter causes stars to move faster than expected and galaxies to cluster in unusual ways. Dark energy appears to be stretching spacetime itself, separating galaxies at ever-increasing speeds. Optical telescopes play a vital role in detecting these effects, whether in the individual explosions of supernovae or the collective light of galaxies. But the process is painstaking, slow and expensive. By letting go of the need for a pretty picture, by switching the graphics from high-res to 8-bit, radio astronomers have levelled up the search for these most elusive and mysterious elements of our Universe. Radio telescopes trace the vast reservoirs of hydrogen, probing the edges of galaxies far beyond their visible limits and outlining the cosmic ruler with impressive efficiency.

Exploring expansive worlds of dark energy and dark matter requires a multiplayer approach. Optical and radio astronomers work together like a well-balanced co-op team: radio astronomers act as the scouts, sketching the rough landscape and identifying major landmarks, while optical astronomers render the details, filling in the scenery with the intricate structures of galaxies and supernovae. Only by combining their strengths can we begin to answer the many outstanding questions about the dark Universe.

There are many scientific quests left to complete before we can say we truly understand dark matter and dark energy. So ready yourself, Player One. The game is far from over.

16

THE DARK SIDE OF THE MOON

> The story of humans is the story of ideas. Scientific ideas that shine light into dark corners.
>
> Jill Tarter, 'Join the SETI Search', February 2009

We have been on quite the journey, tracing our radio waves as they echo through the Universe. Now, having reached the very limits of time and space, it is time to turn around and return home. From the oblivion of the cosmic Dark Ages, through the first stars blazing into existence, we travel along the vast, invisible scaffolding of dark matter, until, at last, we find ourselves back in familiar territory: the Milky Way. Here, we have witnessed the births of planets and searched for signs of alien life. We have listened to the heartbeats of pulsars and followed trails of hydrogen as they spiral towards the galaxy's distant edges. We have revealed asteroids hiding in broad daylight, slipped undetected beneath the defences of Venus' thick clouds and even uncovered ice in the shadowed craters of Mercury.

But as we approach Earth, we find it a changed place. When we launched our exploration of the radio Universe in Chapter 1, it was the 1960s and radio astronomy was still fairly new. It had been only three decades since Karl Jansky built his merry-go-round and two decades since Stanley Hey realised the Sun was jamming the UK's radar defence network. It had taken a bit of time for the physicists and electrical engineers to emerge from the wreckage of the Second World War with their new knowledge of radar and curiosity for the skies. But when they did, there was no stopping them. Day or night, rain or shine, somewhere on Earth radio telescopes were scanning the skies.

I have mentioned the Arecibo 305-m dish many times in this book, because it has been involved in pretty much every area of radio astronomy since its completion in 1963. It was one of the first great radio telescopes, and it remained one of the most powerful radio telescopes in the world, right until its collapse in 2020. The gigantic dish sat still as a statue for just under sixty years, nestled in the Puerto Rican jungle, while both Earth and the skies changed around it. In 1963, the space race was in full swing, a proxy for the ongoing cold war, where dominance was measured by scientific prowess rather than military strength. Dogs, monkeys and humans had been launched into space, orbited Earth and, in some cases, returned safely home. Radio astronomy grew as canny scientists leveraged funding for large radio telescopes by promising military tracking functions. Humanity set its sights on the Moon, landed on its surface and then quickly lost interest, but the skies carried on changing above Arecibo. Satellites became ubiquitous for GPS and communications, and a space station or two began to drift predictably into view. It even witnessed the latest and greatest change to our skies: the launch of mass communication satellites like Starlink, which now trail across the sky in their thousands.

I've read accounts of people walking under the dish in previous decades, accounts that were full of wonder and amazement. I visited the observatory on a pilgrimage in 2022, nominally to attend a workshop, but in a lot of ways to pay my respects to the telescope that had shaped my research field. I drove under the part of the dish that remained, and I felt amazement and wonder, but with a dose of uneasiness added in. Before 2020, when visitors looked up through the large holes in the perforated metal panels above, they saw the dome housing the antenna suspended on eighteen thick steel metal cables attached to three concrete towers about the same height as the Statue of Liberty. In 2020, those cables failed, and the 900-t dome came crashing down, rending the dish below and shattering astronomers' hearts.[1] That great gateway to the radio Universe was permanently closed. When I visited, I stood amongst the ruins, debris still scattered at my feet. I looked up and the antenna was gone. I saw only an uninterrupted expanse of sky, serenely drifting on as if nothing had happened.

Since then, while we have been away wandering the radio Universe, the last remnants of the Arecibo 305-m dish have been cleared, and the first era of radio astronomy has been consigned to history.[2] Yet radio astronomy itself is far from over; sometimes the older generation needs to step aside in order to let the next generation shine. Across the globe, extraordinary telescopes continue to push the boundaries of discovery. Over the past few years, China has completed construction of a colossal 500-m-diameter dish, and by the end of the 2020s, we should see data flowing in from the 130,000 antennas in Australia making up just part of the Square Kilometre Array. I have seen radio telescopes rising from fields, forests, jungles and mountaintops; dishes that dominate skylines; and tiny, solitary antennas floating unnoticed among ducks on a lake. I love to watch them busily scanning the skies in the middle of the day, and to wonder what they might be pointing at. Without the guiding stars of night-time optical astronomy, it's often impossible to tell. And, as this book has hopefully demonstrated, the target of a radio telescope could be *anything*. It could be conversing with planets, but it could also be downloading data taken by satellites orbiting those planets close-up.

When I see the astonishingly detailed optical maps of Mars produced by those orbiters, a small, uncharitable part of me thinks . . . well, that feels like cheating. Get close enough to a planet and of course you'll see it in exquisite detail. From Earth, radar maps will never match the crispness of those optical maps taken from orbit. But what we lose in clarity, we gain in depth, as our radio waves dig deep from a distance, uncovering what is below the surface, whether that be sharp rocks, caves or even water.

Yet the true strength of ground-based astronomy is not in competing with close-range satellite observations but in the longevity it offers. If you recall Arecibo's decades-long radar studies that refined Mercury's rotation rate, or SETI's ongoing, meticulous screening of nearby star systems, this is where radio astronomy shines – patiently building a dataset over time, watching the Universe unfold on timescales beyond a single scientific career, or even a human lifetime.

Mostly, I have used the terms 'ground-based' and 'Earth-based' interchangeably. But as we return to our home planet from the far reaches of the Universe, watching its blue oceans and swirling clouds come into view, we can't help but also notice our constant companion, the Moon. For all of human history, the Moon has been a timekeeper, and a tide-maker. It has been a source of legend and omen, a backdrop to a multitude of love stories, and a trophy at the end of a cold journey amidst a cold war. Soon, the Moon will feel the first human footfall since 1972. The first space age was driven by exploration and by a fierce desire to prove scientific and technological supremacy. This second space age is different. It is about exploration, yes, but also expansion, resources and profit. And radio astronomers are primed to take advantage of humanity's newest plans. Perhaps it is time to extend my definition of 'ground-based' to include the lunar soil that still bears the imprint of Neil Armstrong's first steps.

In the intervening years, the Moon has not changed much. The US flag still stands proud, its permanent, static ripples preserved by the lack of gravity and atmosphere, untouched by rain or wind. Below its constant orbit, Earth and its society of humans have changed immeasurably. Then, there were just a handful of space agencies, all government-funded and only two advanced enough to dominate the space race, namely the US and the USSR. Now, there are over seventy space agencies worldwide, alongside commercially funded companies. Private human space-flight is now a reality, and billionaires can treat space as their personal playground if they choose. Space capsules are now fitted with touchscreens instead of the banks of switches and dials found in a NASA *Apollo* ship. And the crews look a little different too. These days, Pat, Joan and Janet would have a much fairer chance of being onboard themselves, while Neil, Buzz and Mike watched the televised launch from their living rooms with the kids.

An expedition to establish a base on the Moon, or indeed on Mars, will call for a diverse team of people. We'll need radar technicians to scan for safe underground habitats, psychologists to keep everyone sane on the long journeys and biologists to assess the soil for signs of life. Due to the delay in communication between Earth and Mars, we'll need crews that bring a wide range of life and work experiences with them, such that informed decisions can be made quickly, and in isolation from Earth. We'll need geologists, midwives, doctors, chemists, teachers and dentists, and, contrary to Douglas Adams' opinion, we'll need to make sure the spaceship containing the estate agents and hairdressers gets to Mars safely too. And, of course, I would strongly argue we need astronomers as well.

Will those of us who simply want a better look at the stars be welcome on the Moon, amidst those who wish to mine it, settle on it and use it as a springboard to Mars? I do hope so, because as awesome as the new generation of Earth-based telescopes are, Earth has become a noisy place to work.

Every radio telescope must filter out terrestrial signals such as thunderstorms, mobile phones, Starlink satellites, television broadcasts and aeroplanes. These signals are so overwhelming compared to the faint galactic or extragalactic whispers we seek that radio astronomers fiercely protect 'radio-quiet' zones on Earth. Near the Green Bank Telescope, for instance, a federal law forbids radar-sensor automatic doors, enforces wired internet and even has marshals monitoring the environment for rogue transmissions.[3]

On the Moon, none of this would be a concern. There are no atmospheric effects, no thunderstorms, no mobile networks (yet). Just as Grote Reber constructed the world's first telescope dedicated to radio astronomy in his mother's back garden, the Moon – Earth's back garden – is an enticing location for our next great radio telescopes.[4] And, since we are going, we might as well go to the far side, colloquially, and incorrectly, known as the dark side. The Moon is tidally locked, rotating at exactly the same rate as Earth rotates, so from Earth we see the same side at all times. The 'dark side' sees just as much sunlight, though – when we see the New Moon, the far side is entirely illuminated. If we were to place an optical telescope there, it would see the benefit of losing local human-made light-pollution, but the light-pollution from the Sun would remain, keeping optical astronomy a night-shift job. For radio telescopes, however, the sunlight matters little. It may seem a bleak proposition, living on the Moon and never seeing the Earth. But, for all that we might feel adrift and homesick, with our backs towards Earth, the human-made radio pollution is excised along with our view of home. The far side of the Moon will provide the ultimate radio-quiet retreat . . . as long as it doesn't become too popular.

The first preliminary space missions have already begun assessing the radio-quiet conditions of the lunar far side.[5] Future telescopes may be orbiters, taking measurements only while passing above the far side, or autonomous landers unfurling long copper antennas like the jungle vines reclaiming the ruins of Arecibo. Perhaps one day, a vast radio dish will be built, taking advantage of the Moon's weaker gravity to surpass even the scale of our lost legend.[6] We will, theoretically, be able to do everything we can do from Earth, while losing less data to radio frequency interference; we will be able to do it faster, and to a higher quality. In addition, radio observations from the 'dark side' of the Moon are essential to push into the frontier of the 'dark ages' of the Universe. Earth-based radio observations of the Era of the First Stars are imminent, thanks to mega-arrays such as the Square Kilometre Array. But to push back far into the dark ages, to see the Universe in detail before even the stars were formed, it seems there is nowhere on Earth that is quiet enough. I have known colleagues travel to Antarctica and the most remote islands on Earth and still return disappointed. Even there, one can hear too much radio noise to have any hope of detecting such a faint, remote signal as that emanating from the hydrogen just after the Big Bang.

One cannot ignore the opportunity the Moon affords radio astronomy: a clearer, quieter spot from which to observe the Universe and all that it contains, and a chance to push the frontier of looking back in time. I will join my colleagues in planning for this new era of Moon-based astronomy. But you may have noticed by now that I am fiercely protective of the radio dishes scattered around our beautiful planet, just a day's journey away at most, and a journey that does not require one to strap oneself to a missile. There is still so much to uncover from Earth, from home.

There are radio telescopes everywhere on this beautiful planet, listening to the cacophony of mixed messages from outer space. They can't help but listen to each message they receive, learning about the Milky Way, other galaxies, supermassive black holes, pulsars . . . every facet of astronomy. With the Earth's atmosphere little enough of an obstacle, and the ability to simply build a bigger telescope by joining more antennas together, I find it hard to believe we will abandon the fields and deserts of Earth for the craters and highlands of the Moon any time soon.

Whatever the future of radio astronomy holds, we will need to move forward carrying the curiosity of Karl Jansky, the resilience of Grote Reber, the ingenuity of Stanley Hey, the daring of Frank Drake, the tenacity of Jill Tarter and the meticulous eye for detail of Jocelyn Bell Burnell. Radio astronomy will never lose the grit and down-to-earth essence that makes it unique, whether those antennas have the barren lunar landscape, lush Puerto Rican jungle or dusty Martian mountains as their backdrop. And we will never run out of questions, or reasons to tune in to the Universe.

Because wherever you are, you are surrounded by radio waves. Some carry news, some carry messages from loved ones. Some have escaped black holes, some are the echoes from asteroids. Some have even travelled to us from the Era of the First Stars.

It's all there.

We just have to close our eyes, and listen.

A TOURIST'S GUIDE TO RADIO TELESCOPES

Many radio telescopes are open to visitors, and they're genuinely spectacular – vast metal contraptions listening to the Universe with quiet dignity. Unlike optical telescopes, often tucked away beneath protective domes, radio telescopes stand proud in the open, their immense structures visible from miles around. If you ever find yourself travelling nearby, it's well worth making a small detour to witness some real radio science in action.

This isn't a comprehensive list of every radio telescope in the world, but it does include some of the major ones – along with a few personal favourites. Always check online for up-to-date information on opening hours and whether booking is required. These are working scientific facilities, after all, and schedules can change.

One thing to keep in mind: radio telescopes are incredibly sensitive to electromagnetic signals, so mobile phone and Wi-Fi use is often restricted – sometimes for several kilometres. Download your tickets in advance, or bring a printed copy just in case. And if you usually pay with your phone, consider carrying a debit card or a bit of cash, just to be safe.

I'd love to see your holiday pictures! Reach me on email (emma.chapman@nottingham.ac.uk), or on Bluesky (@dreochapman.bsky.social).

The Lovell Telescope, operational since 1957

Jodrell Bank, Goostrey, Cheshire, UK

A fully steerable 76-m radio dish, once the largest in the world. I absolutely love this place: it's set in beautiful grounds and has a brilliant science centre and a gift shop so good that the cashier once asked me if I was sure before charging my card. If you live in England, there's really no excuse – we're a small island! Take the kids, take your friends. You'll probably find me there too.

Website: https://www.jodrellbank.net/

RAF Air Defence Radar Museum

Horning, Norfolk, UK

This is not a working radio telescope, but it is a fascinating museum telling the history of UK air defence from 1936 to the present day. It has models and displays of the radar machinery that laid the foundations for early radio astronomy.

Website: https://www.radarmuseum.co.uk/

LOFAR (Low Frequency Array), operational since 2012

Osdijk Buinen, Netherlands

LOFAR consists of tens of thousands of antennas distributed in clusters across Europe, with a core near Exloo, Netherlands. You can drive up, park and admire these little antennas as they stand statue-still and look back to the dawn of the Universe. This one is probably for the more hardcore radio nerd, as the antennas look much like the ones you might see on your roof, but it's a peaceful place to visit if in the area. Be warned there are no visitor facilities.

Website: https://www.astron.nl/contact/visiting-astron/visit-lofar/

Westerbork Synthesis Radio Telescope (WSRT), operational since 1970

Hooghalen, Netherlands

The Westerbork Synthesis Radio Telescope consists of fourteen 25-m radio telescopes arranged in a line stretching 2.7 km east to west. Ten are fixed in evenly spaced positions, while the remaining four can be moved along rail tracks. Access to the site itself is restricted, but you can park at the nearby Kamp Westerbork and walk quite close up. Prepare yourself for a sobering experience, however, as one cannot ignore the darker heritage of the site as a detention camp, used during the Second World War by the Nazis.

Website: https://www.astron.nl/contact/visiting-astron/visit-westerbork/

Effelsberg Radio Telescope, operational since 1971

Bad Münstereifel, Germany

A 100-m-diameter steerable radio dish that dominates the landscape. There are no guided tours or visits to the interior of the observatory. However there is good access to the public park site (labelled 'Radioteleskop Effelsberg') which leads to the visitors' building after a walk of ten to fifteen minutes.

Website: https://www.mpifr-bonn.mpg.de/effelsberg/visitors

Allen Telescope Array (ATA), operational since 2007

Hat Creek, Shasta County, California, USA

Hidden away amongst snowy mountains and alpine forest, this telescope has to be one of my favourite places on Earth. Part of the Hat Creek Observatory, it is a small outfit, with a tiny visitor centre, so expect just to drive up and go on a self-guided outdoor tour of the array. Well worth an overnight trip if you are in San Francisco, and make sure to check out the beautiful Burney Falls waterfall while you are there.

Website: https://www.seti.org/research/hcro/

Green Bank Observatory, operational since 1958

Green Bank, West Virginia, USA

This site has been home to many telescopes over the years, but since 2000 it has hosted the world's largest fully steerable radio dish, the wonderful 100-m diameter Robert C. Byrd Green Bank Telescope. The observatory grounds are free and open to the public every day from sunrise to sunset, so you can hike along the trails or take a self-guided tour. There are excellent facilities, including a science centre and regular guided tours.

Website: https://greenbankobservatory.org/visit/

Very Large Array (VLA), operational since 1980

Socorro, New Mexico, USA

Twenty-eight radio dishes, each 25 m across, arranged along three vast arms stretching for tens of kilometres across the desert. The VLA is perhaps most recognisable from its starring role in the film *Contact*, based on Carl Sagan's bestselling novel. The site hosts a visitor centre and offers self-guided walking tours among the antennas, along with occasional staff-led guided tours.

Website: https://public.nrao.edu/visit/very-large-array
Virtual tour: https://public.nrao.edu/explore/vla-explorer/

Atacama Large Millimeter/submillimeter Array (ALMA), operational since 2011

Chajnantor Plateau, Atacama Desert, Chile

The Atacama Large Millimeter/submillimeter Array is located 5,000 m above sea level in northern Chile. It consists of sixty-six high-precision radio dishes: fifty-four of them 12 m wide, and twelve smaller ones 7 m across. The antennas can be repositioned across the desert floor using custom-built transporters. In a compact configuration, they span just 160 m; when fully spread out, they cover up to 16 km, allowing astronomers to zoom in on fine cosmic detail or take in the bigger picture. Sadly, the remote site is not easy to visit in person, but a virtual tour is possible.

Website: https://public.nrao.edu/telescopes/alma/
Virtual tour: https://public.nrao.edu/explore/alma-explorer/

FAST (Five-hundred-meter Aperture Spherical Telescope), operational since 2020

Pingtang County, Guizhou Province, China

The Five-hundred-meter Aperture Spherical Telescope – nicknamed *Tianyan* or 'Heaven's Eye' – is nestled in a natural basin in southwestern China. With a dish spanning a staggering 500 m, it holds the title of the world's largest single-dish radio telescope.

The site is equipped for visitors, but it is very remote, so I'd recommend booking your visit through a tour company.

Website: https://fast.bao.ac.cn/

Murriyang, operational since 1961

Parkes, New South Wales, Australia

Parkes Observatory hosts Murriyang, the 64-m radio telescope featured in the film *The Dish*. An icon of Australian science, Murriyang famously supported receiving the *Apollo 11* Moon landing television signals.

Website: https://www.csiro.au/en/about/facilities-collections/ATNF/Parkes-radio-telescope-Murriyang/Visitors-Centre

ACKNOWLEDGEMENTS

First and foremost, my UK editors Georgina Laycock, Katharine Morris, Caroline Westmore and Kate Craigie at John Murray, my US editors Thomas (T. J.) Kelleher, Brian Distelberg and Gillian Sutliff at Basic Books and my literary agent Elizabeth Sheinkman: this book would not have happened without you, and it is so much better for your involvement. Georgina: I remember our first meeting, and how enthusiastic you were about my proposal. I knew then you were the editor for me! You've believed in me through some very hard times; thank you. Kath: I have to say working with you has been incredible. You've made me a better writer with every merciless cut of an over-extended anecdote. You're a motivational genius, and your insight transformed both the manuscript and me as a writer. Sam Wells, copy-editor extraordinaire: you did an incredible job polishing the finished product, but I'll never concede that a slice must be less than half the cake!

I owe a huge debt of gratitude to the multiple astronomical societies that have hosted me throughout my career and tolerated my questions and curiosity about their telescopes. Mansfield and Sutton Astronomical Society and Sherwood Observatory – Steve Wallace, Sparky and 'Cable Rat' to name just a few members – your society is where the passion that originally ignited radio astronomy lives, and it formed a lot of the inspiration for this book.

Thank you to the Royal Society for funding my research and the inflatable planetarium that helps us spread the wonder of astronomy to local schools. To my colleagues at the University of Nottingham: your research inspires me every day, but more valuable than that is the kind, inclusive and intellectually stimulating environment you have all created together. It is by far the best place I have ever worked. Mark Fromhold – your support has been second to none. You have shown me what leadership can be, and it has been wonderful knowing there is someone in power who understands the importance and value of outreach and science communication.

While I have not featured full interviews, I had many interesting conversations during the research phase which helped shape the direction of the book. Prof. Philip Diamond CBE (Director General of the Square Kilometre Array), Dr Sarah Pearce (Director of the SKA-Low Telescope), Dr Wael Farah (Project Scientist for the Allen Telescope Array), Dr Andrew Siemion (Director of Breakthrough Listen), Dr Josh Dillon (HERA research scientist), Prof. Aaron Parsons (Director of the UC Berkeley Radio Astronomy Laboratory), Prof. Jill Tarter (co-founder and retired director of the SETI Institute), Jack Welch (Founding Vice President of the SETI Institute), Prof. Mike Garrett (Director of Jodrell Bank Centre for Astrophysics), Peter East OBE and John Cook (British Astronomical Association Radio Astronomy Section), Prof. Butler Burton (Leiden University), Jay Tate (Spaceguard UK), Prof. Melissa Ness (Columbia University), Prof. Naomi McClure-Griffiths (SKAO Chief Scientist), Prof. Rachel Webster AO (University of Melbourne).

When it comes to family and friends, 'Apologies' feels like a more appropriate heading than 'Acknowledgements' – especially for a second book. With the first, you can thank them for their support as you all discover, together, just how much work writing a book really is. But choosing to do it all again, fully aware of the impact it will have on your time, energy and relationships . . . that's a different matter. At that point, acknowledgements don't quite cover it. So, to my family and friends, I don't just acknowledge, I apologise. I apologise for the eggshells you had to walk on, the times you bit your tongue to stop the words 'you brought this on yourself' from issuing forth. Tim, to paraphrase our favourite singer, 'you showed me wavelengths I can't see with anyone else'. Thank you for being my anchor in the expanse and never losing sight of me even as I drifted out to explore. Lyra, Cassie and Olive, you are each more beautiful and captivating than any star, planet or comet I have ever seen. I love you more than anything in the Universe, to the furthest galaxy and back. Mum, Dad, Rachael, Christine, Sue, Steve, Laura, Becky, Jonathan, Angela, Hannah, Clare, Jennie, Jenny . . . I adore you all . . . is this a good time to tell you I'm writing a third book? (Sorry!)

NOTES

CHAPTER 1: THE MOON

1. R. H. Dicke and R. Beringer, 'Microwave Radiation from the Sun and Moon', *Astrophysical Journal*, 103, 1946, pp. 375–6.
2. J. H. Piddington and H. C. Minnett, 'Microwave Thermal Radiation from the Moon', *Australian Journal of Scientific Research A: Physical Sciences*, 2, 1949, pp. 63–77.
3. 'Borrow the Moon', UK Research and Innovation, https://www.ukri.org/what-we-do/teaching-resources/borrow-the-moon/ (Accessed 25 September 2025).
4. Z. Bay, 'Reflection of Microwaves from the Moon', *Hungarica Acta Physica*, 1, 1947, pp. 1–22; J. H. Dewitt and E. K. Stodola, 'Detection of Radio Signals Reflected from the Moon', *Proceedings of the IRE*, 37(3), 1949, pp. 229–42.
5. 'Andre and Marit's Moonbounce Wedding', https://www.youtube.com/watch?v=RH3z8TwGwrY (Accessed 25 September 2025).
6. B. Koberlein, 'Astronomical Radar: Illuminating Our Understanding of the Solar System', National Radio Astronomy Observatory, 15 December 2021, https://public.nrao.edu/news/astronomical-radar-illuminating-our-understanding-of-the-solar-system/ (Accessed 25 September 2025).
7. L. Carrer et al., 'Radar Evidence of an Accessible Cave Conduit on the Moon Below the Mare Tranquillitatis Pit', *Nature Astronomy*, 8, 2024, pp. 1119–26.

CHAPTER 2: THE SUN

1. J. Bryant, 'Heinrich Hertz's Experiments and Experimental Apparatus: His Discovery of Radio Waves and his Delineation of their Properties' in A. Nordmann, D. Baird and R. I. Hughes (ed.), *Heinrich Hertz: Classical Physicist, Modern Philosopher* (Springer: 1998), pp. 39–58.
2. J. S. Hey, *The Evolution of Radio Astronomy* (Science History Publications: 1973), pp. 14–16.
3. J. S. Hey, 'Solar Radiations in the 4–6 Metre Radio Wave-Length Band', *Nature*, 157(3976), 1946, pp. 47–8.
4. W. M. Goss, *Making Waves: The Story of Ruby Payne-Scott – Australian Pioneer and Radio Astronomer* (Springer: 2013), p. 96.
5. W. Orchiston, P. Robertson and W. T. Sullivan III, *Golden Years of Australian Radio Astronomy: An Illustrated History* (Springer: 2021), p. 116.
6. T. Pulkkinen, 'Space Weather: Terrestrial Perspective', *Living Reviews in Solar Physics*, 4, 2007, article 1.
7. D. M. Oliveira, E. Zesta and K. Garcia-Sage, 'Tracking Reentries of Starlink Satellites During the Rising Phase of Solar Cycle 25', *Frontiers in Astronomy and Space Sciences*, 12, 2025, 1572313. R. A. Marshall, 'Modeling Geomagnetic Induced Currents in Australian Power Networks', *Space Weather*, 15(7), 2017, pp. 895–916.
8. D. Dobrijevic and A. May, 'The Carrington Event: History's Greatest Solar Storm', Space.com, 20 May 2022, https://www.space.com/the-carrington-event (Accessed 19 July 2025).

CHAPTER 3: MERCURY

1. W. E. Howard, A. H. Barrett and F. T. Haddock, 'Measurement of Microwave Radiation from the Planet Mercury', *Astrophysical Journal*, 136, 1962, pp. 995–1004. K. I. Kellermann, '11-cm Observations of the Temperature of Mercury', Nature, 205, 1965, pp. 1091-2.
2. G. H. Pettengill and R. B. Dyce, 'A Radar Determination of the Rotation of the Planet Mercury', *Nature*, 206, 1965, p. 1240.
3. M. A. Slade, B. J. Butler and D. O. Muhleman, 'Mercury Radar Imaging: Evidence for Polar Ice', *Science*, 258(5082), 1992, pp. 635–40.

4. J. K. Harmon et al., 'Radar Mapping of Mercury's Polar Anomalies', *Nature*, 369(6477), 1994, pp. 213–15.
5. G. A. Neumann et al., 'Bright and Dark Polar Deposits on Mercury: Evidence for Surface Volatiles', *Science*, 339(6117), 2013, pp. 296–300.

CHAPTER 4: VENUS

1. C. H. Mayer et al., 'Observations of Venus at 3.15-cm Wave Length', *Astrophysical Journal*, 127, 1958, pp. 1–10.
2. G. H. Pettengill et al., 'A Radar Investigation of Venus', *Astronomical Journal*, 67, 1962, pp. 181–90 (reported after below reference, but accepted to be the first radar detection). A. C. G. Thomson, J. G. Davies and M. Ryle, 'A New Determination of the Solar Parallax by Means of Radar Echoes from Venus', *Nature*, 190(4775), 1961, pp. 519–20.
3. B. G. Kutuza and O. N. Rzhiga, 'Radio Physical Studies of Planets and the Earth at the Institute of Radio Technology and Electronics of the USSR Academy of Sciences' in S. Y. Braude et al. (ed.), *A Brief History of Radio Astronomy in the USSR: A Collection of Scientific Essays* (Springer: 2012), p. 211.
4. R. M. Goldstein and R. L. Carpenter, 'Rotation of Venus: Period Estimated from Radar Measurements', *Science*, 139(3558), 1963, pp. 910–11.
5. J.-L. Margot et al., 'Spin State and Moment of Inertia of Venus', *Nature Astronomy*, 5, 2021, pp. 676–83.
6. T. Stryk, 'Every Picture from Venus' Surface, Ever', *The Planetary Report*, 15 March 2021, https://www.planetary.org/articles/every-picture-from-venus-surface-ever (Accessed 20 July 2025).
7. R. M. Hahn and P. K. Byrne, 'A Morphological and Spatial Analysis of Volcanoes on Venus', *Journal of Geophysical Research: Planets*, 128, 2023, e2023JE007753.
8. NASA/Jet Propulsion Laboratory, *Venus – 3D perspective view of Maat Mons,* NASA Photojournal, 1996, https://photojournal.jpl.nasa.gov/catalog/PIA00106 (Accessed 20 July 2025).

CHAPTER 5: MARS

1. F. Sauro et al., 'Lava Tubes on Earth, Moon and Mars: A Review on Their Size and Morphology Revealed by Comparative Planetology', *Earth Science Reviews*, 209, 2020, p.103288.
2. C. Li et al., 'Layered Subsurface in Utopia Basin of Mars Revealed by Zhurong Rover Radar', *Nature*, 610(7931), 2022, pp. 308–12. R. Orosei et al., 'Radar Evidence of Subglacial Liquid Water on Mars', *Science*, 361(6401), 2018, pp. 490–3.
3. B. Evans, 'How the Beagle 2 Was Lost – Then Found – on Mars', Astronomy.com, 21 December 2023, https://www.astronomy.com/science/how-the-beagle-2-was-lost-then-found-on-mars/ (Accessed 20 July 2025).

CHAPTER 6: THE ASTEROID BELT

1. P. D. Nicholson et al., 'Radar Imaging of Saturn's Rings', *Icarus*, 177(1), 2005, pp. 32–62.
2. For an outstanding overview of the history behind Lovell the man and Lovell the telescope, see J. Bromley-Davenport and B. John, *Space Has No Frontier: The Terrestrial Life and Times of Sir Bernard Lovell* (Bene Factum Publishing: 2013), p. 264.
3. European Space Agency – Near-Earth Objects Coordination Centre, 'Discovery Statistics', ESA NEOCC, https://neo.ssa.esa.int/discovery-statistics (Accessed 20 July 2025).
4. NASA Planetary Defense Coordination Office, *NASA Planetary Defense Strategy & Action Plan,* NASA, 2023, p. 1, https://www.nasa.gov/wp-content/uploads/2023/06/nasa_-_planetary_defense_strategy_-_final-508.pdf (Accessed 20 July 2025).
5. T. Talbert, 'Five Years After the Chelyabinsk Meteor: NASA Leads Efforts in Planetary Defense', NASA, 15 February 2018, https://www.nasa.gov/solar-system/five-years-after-the-chelyabinsk-meteor-nasa-leads-efforts-in-planetary-defense/ (Accessed 20 July 2025).
6. O. P. Popova et al., 'Chelyabinsk Airburst, Damage Assessment, Meteorite Recovery, and Characterization', *Science*, 342(6162), 2013, pp. 1069–73.

7. Jay Tate, interview with the author, 24 September 2021.
8. B. Fu et al., 'Solar Sail Technology: A State of the Art Review', *Progress in Aerospace Sciences*, 86, 2016, pp. 1–19.
9. H. Shen and C. M. Roithmayr, 2015, 'Towing Asteroids With Gravity Tractors Enhanced by Tethers and Solar Sails', presented at the AAS/AIAA Astrodynamics Specialist Conference, Vail, CO, 9–13 August 2015, NASA Langley Research Center, Report No. NF1676L–21158, https://ntrs.nasa.gov/api/citations/20160006420/downloads/20160006420.pdf (Accessed 20 July 2025).
10. P. A. Taylor et al., 'Arecibo Radar Observations of Near-Earth Asteroid (3200) Phaethon During the 2017 Apparition', *Planetary and Space Science*, 167, 2019, pp. 1–8. S. P. Naidu et al., 'Radar Imaging and Physical Characterization of Near-Earth Asteroid (162421) 2000 ET70', *Icarus*, 226(1), 2019, pp. 323–35.

CHAPTER 7: THE MILKY WAY

1. E. C. Stone et al., 'Voyager 1 Observes Low-Energy Galactic Cosmic Rays in a Region Depleted of Heliospheric Ions', *Science*, 341(6142), 2013, pp. 150–3.
2. G. Galilei, *The Sidereal Messenger of Galileo Galilei and a Part of the Preface to Kepler's Dioptrics*, trans. E. S. Carlos (Rivingtons: 1889), Project Gutenberg eBook No. 46036, https://www.gutenberg.org/files/46036/46036-h/46036-h.htm (Accessed 20 July 2025).
3. J. Scott, *The Vinyl Frontier: The Story of NASA's Interstellar Mixtape* (Bloomsbury: 2020), p. 203. (Worth having on the bookshelf just for the title.)
4. J. D. Kraus, 'Grote Reber: Founder of Radio Astronomy', *Journal of the Royal Astronomical Society of Canada*, 82, 1988, pp. 107–14.
5. G. Reber, 'Cosmic Static', *Astrophysical Journal*, 100, 1944, pp. 279–87.
6. R. Frater, W. Goss and H. Wendt, *Four Pillars of Radio Astronomy: Mills, Christiansen, Wild, Bracewell* (Springer: 2017), pp. 67–70.
7. W. N. Christiansen and J. V. Hindman, 'A Preliminary Survey of 1420 Mc/s. Line Emission from Galactic Hydrogen', *Australian Journal of Scientific Research A: Physical Sciences*, 5, 1952, pp. 437–55. H. C. van de Hulst, C. A. Muller and J. H. Oort, 'The Spiral Structure of the

Outer Part of the Galactic System Derived from the Hydrogen Emission at 21 cm Wavelength', *Bulletin of the Astronomical Institutes of the Netherlands*, 12, 1954, pp. 117–19.

8. C. Li et al., 'The Discovery of the Largest Gas Filament in Our Galaxy, or a New Spiral Arm?', *Astrophysical Journal Letters*, 918(1), 2021, L2.
9. M. J. Reid et al., 'Trigonometric Parallaxes of High-Mass Star-Forming Regions: Our View of the Milky Way', *Astrophysical Journal*, 885(2), 2019, pp. 131–49.
10. Alberto Sanna et al., 'Mapping Spiral Structure on the Far Side of the Milky Way', *Science*, 358, 2017, pp. 227–30.
11. I. Heywood et al., 'The 1.28 GHz MeerKAT Galactic Center Mosaic', *Astrophysical Journal*, 925(2), 2022, pp. 165–94.
12. J. Japelj, 'Astonishing Radio View of the Milky Way's Heart', Sky & Telescope, 3 February 2022, https://skyandtelescope.org/astronomy-news/astonishing-radio-view-of-the-milky-ways-heart/ (Accessed 20 July 2025).
13. NASA, *Voyager 1's Pale Blue Dot*, https://science.nasa.gov/mission/voyager/voyager-1s-pale-blue-dot/ (Accessed 20 July 2025).

CHAPTER 8: BLACK HOLES

1. B. P. Abbott et al. (LIGO Scientific Collaboration and Virgo Collaboration), 'Observation of Gravitational Waves from a Binary Black Hole Merger', *Physical Review Letters*, 116(6), 2016, p. 061102.
2. Event Horizon Telescope Collaboration, 'First M87 Event Horizon Telescope results. I. The Shadow of the Supermassive Black Hole', *Astrophysical Journal Letters*, 875(1), 2019, L1.
3. Event Horizon Telescope Collaboration, 'First Sagittarius A* Event Horizon Telescope Results. I. The Shadow of the Supermassive Black Hole in the Center of the Milky Way', *Astrophysical Journal Letters*, 930(2), 2022, L12.
4. See, for example, 'Best-Ever Snapshot of a Black Hole's Jets', Centaurus A (NGC 5128): NASA/Goddard Space Flight Center, 2017, *GSFC_20171208_Archive_e001886,* NASA, https://images.nasa.gov/details/GSFC_20171208_Archive_e001886 (Accessed 20 July 2025).
5. National Radio Astronomy Observatory, *Cygnus A*, NRAO/AUI

Archives, https://www.nrao.edu/archives/items/show/33386 (Accessed 20 July 2025).

6. R. S. Lu et al., 'A Ring-Like Accretion Structure in M87 Connecting Its Black Hole and Jet', *Nature*, 616, 2023, pp. 686–90.

CHAPTER 9: NEUTRON STARS

1. A. Hewish, 'Background to Discovery: Some Recollections', *AIP Conference Proceedings*, 983(1), 2008, pp. 3–7.
2. My favourite account of this time is from Bell Burnell herself, within this exceptional collection of recollections: K. I. Kellermann and B. Sheets (ed.) *Serendipitous Discoveries in Radio Astronomy: Proceedings of a Workshop Held at the National Radio Astronomy Observatory, Green Bank, West Virginia, on May 4, 5, 6, 1983 (Workshop No. 7)* (National Radio Astronomy Observatory: 1983).
3. A. Hewish et al., 'Observation of a Rapidly Pulsating Radio Source', *Nature*, 217, 1968, pp. 709–13.
4. See e.g. C. M. Espinoza et al., 'A Study of 315 Glitches in the Rotation of 102 Pulsars', *Monthly Notices of the Royal Astronomical Society*, 414(2), 2011, pp. 1679–704.
5. S. Tesh and J. Wade, 'Look Happy Dear, You've Just Made a Discovery', *Physics World*, September 2017, pp. 31–3, https://physicsworld.com/a/look-happy-dear-youve-just-made-a-discovery/ (Accessed 20 July 2025).
6. I. Sample, 'British Astrophysicist Overlooked by Nobels Wins $3m Award for Pulsar Work', *Guardian*, 6 September 2018, https://www.theguardian.com/science/2018/sep/06/jocelyn-bell-burnell-british-astrophysicist-overlooked-by-nobels-3m-award-pulsars (Accessed 20 July 2025).
7. Science Museum Group, *Parts from the Cambridge Interplanetary Scintillation Array,* https://collection.sciencemuseumgroup.org.uk/objects/co8105496/parts-from-the-cambridge-interplanetary-scintillation-array (Accessed 20 July 2025).

CHAPTER 10: GRAVITATIONAL WAVES

1. M. Kramer, 'Pulsars as Probes of Gravity and Fundamental Physics', *International Journal of Modern Physics D*, 25(14), 2016, pp. 1630029–61.
2. R. A. Hulse and J. H. Taylor, 'Discovery of a Pulsar in a Binary System', *Astrophysical Journal Letters*, 195, 1975, L51–3.
3. J. H. Taylor and J. M. Weisberg, 'A New Test of General Relativity: Gravitational Radiation and the Binary Pulsar PSR 1913+16', *Astrophysical Journal*, 253, 1982, pp. 908–20.
4. J. M. Weisberg, D. J. Nice and J. H. Taylor, 'Timing Measurements of the Relativistic Binary Pulsar PSR B1913+16', *Astrophysical Journal*, 722(2), 2010, pp. 1030–4.
5. B. P. Abbott, et al. (LIGO Scientific Collaboration and Virgo Collaboration), 'Observation of Gravitational Waves from a Binary Black Hole Merger', *Physical Review Letters*, 116(6), 2016, p. 061102.
6. G. Agazie et al. (NANOGrav Collaboration), 'The NANOGrav 15 yr Data Set: Evidence for a Gravitational-Wave Background', *Astrophysical Journal Letters*, *951*(1), 2023, L8.
7. L. Chiesa, A. Perego and F. M. Guercilena, 'Did a Kilonova Set Off in Our Galactic Backyard 3.5 Myr Ago?', *Astrophysical Journal Letters*, 962(2), 2024, L24.

CHAPTER 11: EXOPLANETS

1. M. Mayor, and D. Queloz, 'A Jupiter-Mass Companion to a Solar-Type Star', *Nature*, 378, 1995, pp. 355–9.
2. A. Wolszczan and D. A. Frail, 'A Planetary System Around the Millisecond Pulsar PSR1257+12', *Nature*, 355(6356), 1992, pp. 145–7.
3. 'Exoplanets', NASA, https://science.nasa.gov/exoplanets/ (Accessed 20 July 2025).
4. ALMA Partnership, et al., 'The 2014 ALMA Long Baseline Campaign: First Results from High Angular Resolution Observations Toward the HL Tau Region', *Astrophysical Journal Letters* 808(1), 2015, L3.
5. National Radio Astronomy Observatory, *Twenty Protoplanetary Disks Imaged by ALMA*, NRAO Public Gallery, https://public.nrao.edu/

gallery/twenty-protoplanetary-disks-imaged-by-alma/ (Accessed 20 July 2025).

6. S. Facchini et al., 'Resolved ALMA Observations of Water in the Inner Astronomical Units of the HL Tau Disk', *Nature Astronomy*, 8, 2024, pp. 587–95.
7. E. Herbst and E. F. van Dishoeck, 'Complex Organic Interstellar Molecules', *Annual Review of Astronomy and Astrophysics*, 47(1), 2009, pp. 427–80.
8. Center for Astrophysics Harvard & Smithsonian, 'First Detection of PreBiotic Molecule Glycolonitrile in Space', Harvard Center for Astrophysics News, 8 March 2019, https://www.cfa.harvard.edu/news/first-detection-pre-biotic-molecule-glycolonitrile-space (Accessed 20 July 2025).

CHAPTER 12: CONTACT

1. M. Gillon et al., 'Seven Temperate Terrestrial Planets Around the Nearby Ultracool Dwarf Star TRAPPIST−1', *Nature*, 542(7642), 2017. pp. 456–60.
2. M. Brown, *How I Killed Pluto and Why It Had It Coming* (Spiegel & Grau: 2012), p. vii.
3. P. D. Ward and D. Brownlee, *Rare Earth: Why Complex Life Is Uncommon in the Universe* (Springer: 2004).
4. W. Arthur, *The Biological Universe: Life in the Milky Way and Beyond* (Cambridge University Press: 2020).
5. F. D. Drake, 'Project Ozma', *Physics Today,* 14(4), 1 April 1961, pp. 40–6.
6. F. D. Drake and D. Sobel, *Is Anyone Out There? The Scientific Search for Extraterrestrial Intelligence* (Delacorte Press: 1994), p. 46. (In the author's opinion, one of the best popular science books ever written.)
7. G. R. Harp et al., 'SETI Observations of Exoplanets with the Allen Telescope Array', *Astronomical Journal*, 152(6), 2016, pp. 181–94.
8. S. Bowyer et al., 'The Berkeley Parasitic SETI Program', *Icarus*, 53(1), 1983, pp. 147–55.
9. S. Scoles, *Making Contact: Jill Tarter and the Search for Extraterrestrial Intelligence* (Pegasus: 2017), p. 4.

10. 'Initiative Overview: Breakthrough Listen', Breakthrough Initiatives, https://breakthroughinitiatives.org/initiative/1 (Accessed 20 July 2025). N. Franz et al., 'The Breakthrough Listen Search for Intelligent Life: Technosignature Search of Transiting TESS Targets of Interest', *Astronomical Journal*, 163(3), 2022, pp. 104–14.
11. N. Drake, 'The Arecibo Message, Earth's First Interstellar Transmission, Turns 50', *Scientific American*, 14 November 2024, https://www.scientificamerican.com/article/the-arecibo-message-earths-first-interstellar-transmission-turns-50/ (Accessed 20 July 2025).
12. W. Roush, *Extraterrestrials* (MIT Press: 2020), p. 62.

CHAPTER 13: FIRST LIGHT

1. ESO Supernova, Bell Labs Horn Antenna, Crawford Hill, New Jersey, 1964, https://supernova.eso.org/exhibition/images/penzias-wilson-hornantenna-cc/ (Accessed 20 July 2025).
2. Nature Video, *The Pigeon, the Antenna and Me: Robert Wilson*, *Scientific American*, 21 October 2015, https://www.scientificamerican.com/video/the-pigeon-the-antenna-and-me-robert-wilson/ (Accessed 21 July 2025).
3. R. Schoenstein, 'The Big Bang's Echo', *All Things Considered*, NPR, 17 May 2005, https://www.npr.org/2005/05/17/4655517/the-big-bangs-echo (Accessed 21 July 2025).
4. R. H. Dicke et al., 'Cosmic Black-Body Radiation', *Astrophysical Journal*, 142, 1965, pp. 414–19.
5. A. A. Penzias and R. W. Wilson, 'A Measurement of Excess Antenna Temperature at 4080 Mc/s.', *Astrophysical Journal*, 142, 1965, pp. 419–21.
6. There are plenty of books on the Big Bang (e.g. Simon Singh's classic, *Big Bang: The Origin of the Universe* (Fourth Estate: 2005)), but this is one of the more thoughtful, if sometimes bizarre, contributions outlining exactly what we do know, and what we don't: N. Afshordi, and P. Halper, *Battle of the Big Bang: The New Tales of Our Cosmic Origins* (University of Chicago Press: 2025).
7. E. Hubble, 'A Relation Between Distance and Radial Velocity Among Extra-Galactic Nebulae', *Proceedings of the National Academy of Sciences of the United States of America*, 15(3), 1929, pp. 168–73.

8. C. O'Raifeartaigh and S. Mitton, 'Interrogating the Legend of Einstein's "Biggest Blunder"', *Physics in Perspective*, 20(4), 2018, pp. 318–41.
9. A. G. Lemaître, 'A Homogeneous Universe of Constant Mass and Increasing Radius Accounting for the Radial Velocity of Extra-Galactic Nebulae', *Monthly Notices of the Royal Astronomical Society*, 91, 1931, pp. 483–90.
10. P. Halpern, *Flashes of Creation: George Gamow, Fred Hoyle, and the Great Big Bang Debate* (Basic Books: 2021).
11. M. Schmidt, '3C 273: A Star-Like Object with Large Red-Shift', *Nature*, 197(4872), 1963, p. 1040.
12. I. Kellermann, 'The Discovery of Quasars', *Bulletin of the Astronomical Society of India*, 41(1), 2013, pp. 1–17.
13. K. Cooper, 'Quasars: Brightest Objects in the Universe', Space.com, 19 October 2023, https://www.space.com/17262-quasar-definition.html (Accessed 21 July 2025).
14. S. Weinberg, *The First Three Minutes: A Modern View of the Origin of the Universe* (Basic Books: 1993).
15. E. Chapman, *First Light: Switching on Stars at the Dawn of Time* (Bloomsbury: 2022).
16. E. Chapman, 'First Light: Switching on Stars at the Dawn of Time', *Contemporary Physics*, 63(1), 2022, pp. 15–33. (A more scientific review than the above popular science book.)
17. R. S. Klessen et al., 'The First Stars: Formation, Properties, and Impact', *Annual Review of Astronomy and Astrophysics*, 61, 2023, pp. 65–130.
18. See, e.g. K. Cooper, 'James Webb Space Telescope Finds 2 of the Most Distant Galaxies Ever Seen', Space.com, 14 November 2023, https://www.space.com/james-webb-space-telescope-distant-galaxies (Accessed 21 July 2025).
19. SKAO (Square Kilometre Array Observatory), 2025, https://www.skao.int/en (Accessed 21 July 2025).

CHAPTER 14: DARK MATTER

1. H. I. Ewen and E. M. Purcell, 'Observation of a Line in the Galactic Radio Spectrum: Radiation from Galactic Hydrogen at 1,420 Mc./sec.', *Nature*, 168(4270), 1951, p. 356.
2. G. Byrd and M. Valtonen, 'How Radio Tuned in to Dark Matter', *Astronomy & Geophysics*, 61(6), December 2020, pp. 6.26–9.
3. H. C. van de Hulst, E. Raimond and H. van Woerden, 'Rotation and Density Distribution of the Andromeda Nebula Derived from Observations of the 21-cm Line', *Bulletin of the Astronomical Institutes of the Netherlands*, 14, 1957, pp. 1–16.
4. W. J. G. de Blok, et al., 'High-Resolution Rotation Curves and Galaxy Mass Models from THINGS', *Astronomical Journal*, 136(6), 2008, pp. 2648–719.
5. To visualise the dark matter web, lose yourself in this cosmological simulation: Illustris Collaboration, https://www.illustris-project.org/media/

CHAPTER 15: DARK ENERGY

1. A. G. Riess et al., 'Observational Evidence from Supernovae for an Accelerating Universe and a Cosmological Constant', *Astronomical Journal*, 116(3), 1998, pp. 1009–38. S. Perlmutter et al., 'Discovery of a Supernova Explosion at Half the Age of the Universe', *Nature*, 391, 1998, pp. 51–4.
2. NASA/Goddard Space Flight Center, 'Baryon Acoustic Oscillations', NASA Scientific Visualization Studio, 18 November 2020, https://svs.gsfc.nasa.gov/13768 (Accessed 21 July 2025).
3. National Radio Astronomy Observatory, 'The Mysterious Dark Energy', https://public.nrao.edu/radio-astronomy/dark-energy/ (Accessed 21 July 2025).
4. M. Santos et al., 'Cosmology from a SKA HI Intensity Mapping Survey', *Advancing Astrophysics with the Square Kilometre Array (AASKA14)*, 19, 2015. P.-J. Wu et al., 'Prospects for Measuring Dark Energy with 21 cm Intensity Mapping Experiments: A Joint Survey Strategy', *Science China Physics, Mechanics, and Astronomy*, 66, 2023, p. 270413.

5. S. Cunnington et al., 'H I Intensity Mapping with MeerKAT: Power Spectrum Detection in Cross-Correlation with WiggleZ Galaxies', *Monthly Notices of the Royal Astronomical Society,* 518, 2023, pp. 6262–72. T.-C. Chang et al., 'An Intensity Map of Hydrogen 21-cm Emission at Redshift z ~ 0.8', *Nature,* 466, 2010, pp. 463–5.
6. K. Mack, *The End of Everything: (Astrophysically Speaking)* (Allen Lane: 2020).

CHAPTER 16: THE DARK SIDE OF THE MOON

1. National Science Foundation (NSF), *Footage of Arecibo Observatory Telescope Collapse,* 2020, https://youtu.be/ssHkMWcGat4 (Accessed 21 July 2025).
2. C. Feehly, '4 Years After the Giant Arecibo Observatory Collapsed, We Finally Know What Happened', Space.com, 8 November 2024, https://www.space.com/the-universe/4-years-after-the-giant-arecibo-observatory-collapsed-we-finally-know-what-happened (Accessed 21 July 2025).
3. S. Kurczy, *The Quiet Zone: Unraveling the Mystery of a Town Suspended in Silence* (HarperCollins: 2021), p. 49.
4. J. O. Burns, 'Transformative Science from the Lunar Farside: Observations of the Dark Ages and Exoplanetary Systems at Low Radio Frequencies', *Philosophical Transactions of the Royal Society A*, 379, 2021, p. 20190564.
5. C. Li et al., 'Overview of the Chang'e-4 Mission: Opening the frontier of scientific exploration of the Lunar Far Side', *Space Science Reviews,* 217, 2021, p. 35.
6. M. Hutson, 'The Quest to Build a Telescope on the Moon', *New Yorker,* 24 September 2024, https://www.newyorker.com/science/elements/the-quest-to-build-a-telescope-on-the-moon (Accessed 21 July 2025).

FURTHER READING

GENERAL

Baars, J. W. M. and H. J. Kärcher, *Radio Telescope Reflectors: Historical Development of Design and Construction* (Springer: 2017)

Burke, B. F. and F. Graham-Smith, *An Introduction to Radio Astronomy* (Cambridge University Press: 2002) (The definitive textbook for the mathematically curious.)

Fielding, J., *Amateur Radio Astronomy* (Radio Society of Great Britain: 2010)

Graham-Smith, F., *Eyes on the Sky: A Spectrum of Telescopes* (Oxford University Press: 2016)

Kellermann, K. I. and B. Sheets, *Serendipitous Discoveries in Radio Astronomy: Proceedings of a Workshop Held at the National Radio Astronomy Observatory, Green Bank, West Virginia on May 4, 5, 6, 1983: Workshop No. 7* (National Radio Astronomy Observatory: 1983) (Simply the best resource for first-hand accounts of every major discovery in radio astronomy in the early decades.)

Lashley, J., *The Radio Sky and How to Observe It* (Springer: 2010)

Lovell, B., *Voice of the Universe: Building the Jodrell Bank Telescope* (Praeger: 1987)

Parachini, J., *Listening to the Stars: Jocelyn Bell Burnell Discovers*

Pulsars (Albert Whitman: 2021) (This is a children's book, but it is awesome.)

Romero, R. *The Reception and Study of Naturally Originating Radio Signals* (Radio Society of Great Britain: 2008)

Sullivan, W. T., *Classics in Radio Astronomy* (Springer: 2012) (Contains many of the first scientific papers in radio astronomy.)

Sullivan, W. T., *Cosmic Noise: A History of Early Radio Astronomy* (Cambridge University Press: 2009) (An excellent, if dense, record of the very first years of radio astronomy, including an in-depth look at Grote Reber.)

Wild, S., *Searching African Skies* (Jacana Media: 2012)

Yong, E., *An Immense World: How Animal Senses Reveal the Hidden Realms Around Us* (Bodley Head: 2022) (An excellent reminder of how limited humans are in our perception of light.)

CHAPTER 1: THE MOON

Tyson, N. D. and A. Lang, *Accessory to War: The Unspoken Alliance Between Astrophysics and the Military* (W. W. Norton: 2018)

For a flavour of the ambition and adrenaline involved in the US space race, I can highly recommend:

Chaikin, A., *A Man on the Moon: The Voyages of the Apollo Astronauts* (Penguin: 2019)

Kranz, G., *Failure Is Not an Option: Mission Control from Mercury to Apollo 13 and Beyond* (Simon & Schuster: 2000)

Scott, D. and A. A. Leonov, *Two Sides of the Moon: Our Story of the Cold War Space Race* (Macmillan: 2004)

Smith, A., *Moondust: In Search of the Men Who Fell to Earth* (Updated edn) (Bloomsbury: 2019)

CHAPTER 2: THE SUN

Bartlett, R. A., *The World of Ham Radio, 1901–1950: A Social History* (McFarland: 2015)
Braude, S. Y. et al., *A Brief History of Radio Astronomy in the USSR: A Collection of Scientific Essays* (Springer: 2012)
Cohen, R., *Chasing the Sun: The Epic Story of the Star That Gives Us Life* (Simon & Schuster: 2011)
Frater, R., W. Goss and H. Wendt, *Four Pillars of Radio Astronomy: Mills, Christiansen, Wild, Bracewell* (Springer: 2017)
Green, P. L., *15 Million Degrees: A Journey to the Centre of the Sun* (Viking: 2016)

CHAPTER 4: VENUS

Greenspoon, D. H., *Venus Revealed: A New Look Below the Clouds of Our Mysterious Twin Planet* (Basic Books: 1998)
Sobel, D., *The Planets* (Viking: 2005) (A beautiful account of our Solar System.)

CHAPTER 5: MARS

Bradbury, R., *The Martian Chronicles* (Simon & Schuster: 2012) (Nothing to do with the science of this chapter, but a book you should have on your shelf.)
Weinersmith, K. and Z. Weinersmith, *A City on Mars: Can We Settle Space, Should We Settle Space, and Have We Really Thought This Through?* (Particular Books: 2023)

CHAPTER 6: THE ASTEROID BELT

Gordon, H., *The Meteorites: Encounters with Outer Space and Deep Time* (Profile: 2025)

Howgego, J., *The Meteorite Hunters: On the Trail of Extraterrestrial Treasures and the Secrets Inside Them* (Oneworld: 2025)

CHAPTER 8: BLACK HOLES

Bartusiak, M., *Black Hole: How an Idea Abandoned by Newtonians, Hated by Einstein, and Gambled on by Hawking Became Loved* (Yale University Press: 2015)

Falcke, H. and J. Römer, *Light in the Darkness: Black Holes, the Universe and Us* (Wildfire: 2021)

CHAPTER 9: NEUTRON STARS

Moskvitch, K., *Neutron Stars: The Quest to Understand the Zombies of the Cosmos* (Harvard University Press: 2020)

CHAPTER 10: GRAVITATIONAL WAVES

Lyne, A., F. Graham-Smith and B. Stappers, *Pulsar Astronomy* (Cambridge University Press: 2022)

CHAPTER 11: EXOPLANETS

Tasker, E., *The Planet Factory: Exoplanets and the Search for a Second Earth* (Bloomsbury: 2019)

CHAPTER 12: CONTACT

Darling, D., *Life Everywhere* (Basic Books: 2007)

CHAPTER 13: FIRST LIGHT

Gastrow, M., *The Stars in Our Eyes: Representations of the Square Kilometre Array Telescope in the South African Media* (HSRC Press: 2017)

Schilizzi, R. T. et al., *The Square Kilometre Array: A Science Mega-Project in the Making, 1990–2012* (Springer: 2024)

CHAPTER 14: DARK MATTER

Clegg, B., *Dark Matter and Dark Energy: The Hidden 95% of the Universe* (Icon: 2019)

Yeager, A. J., *Bright Galaxies, Dark Matter, and Beyond: The Life of Astronomer Vera Rubin* (MIT Press: 2023)

CHAPTER 16: THE DARK SIDE OF THE MOON

Nan, R., *The Sky Eye: Five-Hundred-Meter Aperture Spherical Radio Telescope (FAST)* (Springer: 2021)

INDEX

INDEX

Photo Credit: Dan Wray

Emma Chapman is an award-winning astrophysicist and Royal Society Research Fellow based at the University of Nottingham. She is among the world's leading researchers in the search for the first stars of our universe and is the author of *First Light: Switching on Stars at the Dawn of Time*. She lives in Nottinghamshire, UK.

RAISING READERS

Books Build Bright Futures

Thank you for reading this book and for being a reader of books in general. We are so grateful to share being part of a community of readers with you, and we hope you will join us in passing our love of books on to the next generation of readers.

Did you know that reading for enjoyment is the single biggest predictor of a child's future happiness and success?

More than family circumstances, parents' educational background, or income, reading impacts a child's future academic performance, emotional well-being, communication skills, economic security, ambition, and happiness.

Studies show that kids reading for enjoyment in the US is in rapid decline:

- In 2012, 53% of 9-year-olds read almost every day. Just 10 years later, in 2022, the number had fallen to 39%.
- In 2012, 27% of 13-year-olds read for fun daily. By 2023, that number was just 14%.

TOGETHER, WE CAN COMMIT TO RAISING READERS AND CHANGE THIS TREND. HOW?

- Read to children in your life daily.
- Model reading as a fun activity.
- Reduce screen time.
- Start a family, school, or community book club.
- Visit bookstores and libraries regularly.
- Listen to audiobooks.
- Read the book before you see the movie.
- Encourage your child to read aloud to a pet or stuffed animal.
- Give books as gifts.
- Donate books to families and communities in need.

Books build bright futures, and **Raising Readers** is our shared responsibility.

For more information, visit JoinRaisingReaders.com

Sources: National Endowment for the Arts, National Assessment of Educational Progress, WorldBookDay.com, Nielsen BookData's 2023 "Understanding the Children's Book Consumer"